Stress and Animal Welfare

D. M. Broom

Department of Clinical Veterinary Medicine
University of Cambridge
UK

and

K.G. Johnson

School of Veterinary Studies
Murdoch University
Australia

CHAPMAN & HALL
London · Glasgow · New York · Tokyo · Melbourne · Madras

Published by Chapman & Hall, 2–6 Boundary Row, London SE1 8HN

Chapman & Hall, 2–6 Boundary Row, London SE1 8HN, UK

Blackie Academic & Professional, Wester Cleddens Road, Bishopbriggs, Glasgow G64 2NZ, UK

Chapman & Hall Inc., One Penn Plaza, 41st floor, New York, NY 10119, USA

Chapman & Hall Japan, Thomson Publishing Japan, Hirakawacho Nemoto Building, 6F, 1-7-11 Hirakawa-cho, Chiyoda-ku, Tokyo 102, Japan

Chapman & Hall Australia, Thomas Nelson Australia, 102 Dodds Street, South Melbourne, Victoria 3205, Australia

Chapman & Hall India, R. Seshadri, 32 Second Main Road, CIT East, Madras 600 035, India

First edition 1993

Typeset in 11/12 pt Bembo by EXPO Holdings, Malaysia
Printed in Great Britain by TJ Press, Padstow

ISBN 0 412 39580 0

A catalogue record for this book is available from the British Library

Library of Congress Cataloging-in-Publication data

Broom, Donald M.
Stress and animal welfare / D. Broom and K.G. Johnson. – 1st ed.
p. cm. – (Chapman and Hall animal behaviour series)
Includes bibliographical references and index.
ISBN 0–412–39580–0 (alk. paper)
1. Stress (Physiology) 2. Animal welfare. I. Johnson, K.G.
II. Series
QP82.2.S8B76 1993
591.1'88–dc20

93-32185
CIP

Printed on permanent acid-free text paper, manufactured in accordance with the proposed ANSI/NISO Z 39.48-199X and ANSI Z 39.48-1984

Contents

Preface

Developing a new scientific discipline and making it applicable to everyday life depends upon establishing certain key concepts and evolving skills in using them. Key concepts provide bases from which deductions can be made, and structures to which emerging ideas can be attached. Since the upsurge of public interest in the welfare of animals, there has been much study of **welfare** and of **stress**, to which welfare is obviously related. This interest has extended across many scientific domains, from human medicine, animal biology, veterinary medicine and agriculture to psychology and philosophy.

The significance of the terms stress and welfare has stimulated a considerable amount of writing and no small amount of debate, especially because of their relevance to humans and domestic animals. We are now at a stage where there is a pressing need to build the terms into a conceptual scheme that is sufficiently sound to provide predictions, to allow new data to be added, and to be subjected to exacting tests.

From first principles, we believe there is no reason why the concepts of stress and welfare should be essentially different whether used for humans or for other animals, so the ideas developed will be structured to refer to all animals, both human and non-human. The examples and emphases will nonetheless principally relate to non-human animals, since that is the focus of this series of texts and the background of the authors.

In Chapter 1, the need for careful scientific study of stress and welfare is explained. The reasons for some of the problems in understanding the concepts are discussed, and it is argued that there is a requirement for further analysis of the concepts, and especially for a better synthesis of current ideas. We seek to clarify use of the terms stress and welfare by deriving definitions for them related to the functioning and efficacy of the biological systems that animals use to both regulate their lives and deal with difficulties. These systems include a wide range of biological components including the feelings of the animals. This derivation is explained in Chapters 2 and 3. The definitions, based on established biological concepts and consistent with similar ideas in other disciplines, are described in detail in Chapter 4.

From this theoretical base, sound and practical approaches for assessing welfare are outlined. Chapter 5 provides an account of the responses of animals to short-term disturbances, while the responses to long-term disturbances are documented in Chapter 6. In Chapter 7 the use of animal preference studies to provide information relevant to the assessment of animal welfare is discussed. The question of how great a disturbance of homeostasis, or what level of stimulation an animal should be subjected to is partly a matter of biological judgement, since animals may manage better if exposed to a moderate level of stimulation, even if it is aversive, rather than being protected from stimulation entirely. But ethical considerations obviously also dictate that there must be a limit. A survey of the ethical issues involved and a guide to making ethical decisions about animal stress and welfare are presented in Chapter 8.

Finally, what is believed to be the best current approach for monitoring animal welfare is outlined in Chapter 9. It advocates combining technical measures of stress and welfare, based on a sound biological framework, with appropriate ethical limitations.

Evolution of human society is constantly changing the relationship between humans and other animals, but too often this is to the detriment of those animals. Fortunately, biological studies are uncovering ways of identifying, assessing and alleviating poor welfare. With this information, strategies can be developed to avoid unreasonable impositions on animals. The ultimate goal of the book is to establish a biological base from which can be developed codes of animal management appropriate to a modern and compassionate society.

Acknowledgements

We thank C. E. Manser, G. J. Mason, M. T. Mendl and other members of the Animal Welfare Discussion Group in Cambridge for contributions to the collection of material, help in formulation of ideas and comments on the manuscript for this book. We also thank E. Kirby, J. Blackburn, S. E. M. Broom and the editorial staff at Chapman & Hall for help in preparation of the book.

Chapter 1
Approaching questions of stress and welfare

1.1 THE RANGE OF USE OF THE TERMS STRESS AND WELFARE

Stress and welfare are words found in many current newspapers, magazines and novels, as well as in numerous academic journals and texts. Because they are widely used, the terms stress and welfare have developed a range of meanings. The terms tend to be used in order to avoid being too specific about the nature of particular difficulties. Many of the issues surrounding these difficulties are receiving increasing public and scientific attention because of their social and ethical implications. Unfortunately, the more widely the words are used, the less precise are the meanings attached to them.

The terms stress and welfare strike responsive chords in all of us, perhaps because they relate to our concerns about our own well-being, including our health, and potential threats to that well-being. We have a vested interest in the stresses imposed on us, and their effects on our welfare. The terms are also key components of the public debate about animal welfare. An extension of our worries about the stresses imposed on us is a concern about the stresses put on animals that we live with, or that are in our charge. Disquiet about these has prompted challenges to various types of animal husbandry practised by our society.

This book develops ideas that could be applicable to all species, including humans. Special efforts have been made to include references to human data in order to make the concepts as general as possible. There has been an enormous amount of scientific and social study of human stress and welfare and, although it is intended that the theories offered here should be applicable to humans, only a small fraction of the vast literature on human stress and welfare is cited.

The principal emphasis in the book is therefore on non-humans and, amongst them, on the more commonly seen mammals and birds, since these are the species in which stress and welfare have been most studied and with which the authors are most familiar. In reviewing such studies, we have borne in mind that stress and welfare in various managed or exploited animals need to be understood before animal welfare requirements can be prescribed in legislation.

1.2 ANIMAL WELFARE AND SOCIAL CHANGE

The rise in public interest in animal welfare during the past two decades has been dramatic. Concern for animals is evident throughout society in many countries and is invisible only to those who do not want to see. Attention has been drawn, by both responsible and irresponsible means, to the various ways in which humans interact with animals. Particularly loud have been claims that some or all of what is being done to animals is unjustified. The wide disparity in people's views on how humans should treat animals has led to social polarization in a debate about the morality of various human–animal interactions. Why did this debate emerge? Why has it evoked such strong feelings? Why is there special interest in it now? An examination of the controversy may help us understand its significance and contribute to resolving it.

Human–animal interactions must be as old as mankind. From earliest times, simple, one-sided interactions have occurred when animals are killed by man to supply a resource of some kind, commonly food. Conversely humans must also have been killed by animals. Humans have, for thousands of years, managed populations of animals for transport, recreation, companionship or protection. For our convenience and ease of management animals have in recent times been kept in or near human dwellings, and their genotype modified by selective breeding.

Domesticated animals are certainly exploited by man, but to some extent they have also exploited the human ecosystem (Budiansky, 1992). The domestic fowl, by providing a food source for man and adapting to human conditions, has so encouraged human care that it has become the commonest bird in the world (Broom, 1986a). According to some biological definitions, it has undeniably been a success. The chicken has exploited an ecological niche, a major component of which is the availability of human care. As far as individual chickens are concerned, however, the majority of animals derive much less benefit from humans, in terms of the quality of their lives, than humans derive from them. Food animals, pet animals, and animals that race for human entertainment may be common, and arguably biologically successful, but undeniably all are exploited to some extent within human society.

With passing centuries, the nature of many human–animal interactions has changed. Some activities involving animals, that were once forms of daily business, such as riding horses and driving pony traps, are now competitive sports. Others, like poultry raising, that were previously part of life on an average farm, and involved daily contact with, and knowledge of, each hen or cockerel, are now aggregated into commercial enterprises in which there is little concern for individual animals. Even the keeping of family pets is affected by fashion and business. Animals are bred, purely as a human whim, with characteristics of supposed cosmetic appeal that impair their ability to cope with everyday life.

Human attitudes to animals also change with time, as do other characteristics of an evolving society. In one era people derive enjoyment from the spectacle of animal fights, in another they collect animals in zoos or museums, and in another, with the aid of technology, they seek to maximize animal production and use animals to find cures for diseases. Perceived values of human and animal life are undoubtedly major determinants of what is acceptable. Sections of society in one generation attach great significance to hunting wild animals and in another to conserving them in their natural state. As ideological and economic priorities change, so does society's regard or disregard for animals. But are there absolute standards to which these attitudes should conform? In the current animal welfare debate, some attempt is being made to answer this question.

1.3 THE CURRENT DEBATE ABOUT ANIMAL USAGE

In the present era, social attitudes have been challenged by claims that the interests of non-human animals are being neglected, and that we have a moral obligation to consider them more than we do at present. In order to be able to respond rationally to these concerns, two specific steps are essential. The first is to formulate an ethically and scientifically defensible philosophy about animals. The second is to develop specific animal management practices consistent with this fundamental philosophy. Interested people have found it relatively easy to identify areas for criticisms about current animal usage, but they have been far less successful in framing a convincing philosophy about animals, and using it as a basis for practical solutions to welfare problems.

There are two obvious reasons for the difficulties. Although some scientific information and some ethical propositions about the place of animals in the world have been collected, there is still much we do not know. Secondly, even with such information, there are no simple answers. There seem to be no absolute 'rights' and few certain 'wrongs'. The community faces a classical dilemma. Any solution involves some gains and some losses. These problems are compounded by ignorance about our own species. How distressing are different sorts of human suffering? How can people regulate their lives so as to minimize adverse environmental effects? Should we avoid all potentially harmful stimuli, or is that unwise? An efficient and optimistic approach would be to realize that solutions to human and animal welfare problems can be sought simultaneously through largely similar methods of study. Many biological systems are common to both humans and animals with which humans interact. We can learn about human systems for coping with adversity by studying similar systems in other species. Likewise, many ways of improving human welfare are undoubtedly appropriate for the animals that we keep.

Contributions to the current debate about animal usage, animal rights and animal welfare come from various community groups, ranging from critics who oppose virtually all animal practices to self-confident defenders of current practices or even of undoubted animal exploitation. Some of those espousing extreme positions are philosophers and biologists, but more moderate positions are also adopted by people from these disciplines. Each group has a reason for speaking and a contribution to make. Regrettably, the groups rarely discuss concepts with one another.

Most prominent in the debate are the vocal critics who vigorously denounce society's attitude to animals. Their views vary from beliefs that humans should try to live in such a way that no animal is ever killed, to more practical commitments to finding homes for stray cats. Leaving aside those who will not listen to the views of others, there are in this group many who are constructive in their proposals. They contribute to the debate by identifying areas of concern, and indicating the extent of disquiet within the community. Their role in lobby groups has dramatically influenced the course of animal welfare politics.

Defenders of the present situation with regard to the ways in which animals are treated generally insist that humans should always come before animals. This insistence is not always convincing, especially when some people, even some of these defenders, keep pets in better conditions than humans are forced to endure elsewhere in the world. Furthermore, many people are unaware of the worst conditions in which animals are kept. Some do not know, or choose not to think about, the ways in which animals' conditions have been so strikingly altered from the relatively tranquil conditions that often prevailed during their own or their parents' childhood. For many people in this group, to hear an account of the less savoury current animal practices may well be unsettling. But others in this group know about the science and ethics of animal welfare, and are convinced that, as long as care is taken, present practices are acceptable.

Another group of contributors are the philosophers, of whom Singer (e.g. 1990), Regan (e.g. 1983) and Rollin (e.g. 1981) are perhaps the best known. The conceptual analyses and soul-searching of these thinkers have clarified some of the fundamental issues. However, the theoretical problems have never been entirely solved. There remains a striking lack of unanimity about the philosophical basis of human concern for animals (Miller and Williams, 1983). Amongst different nationalities and religions, philosophical positions concerning animals vary greatly. Representatives from almost all religions make statements urging that the welfare of animals be considered, but practising members of those religions adopt a wide range of attitudes about what that entails. Finally, philosophical analysis must be reconciled with biological reality; between philosophy and biology there are currently major disparities in approach, for example: in emphasizing the importance of rights with little consideration of welfare, in understanding the biological significance of pain (Chapter 5) and in appreciating the distribution of sentience in the animal kingdom. Philosophical enquiry about animal welfare will continue, but the likelihood of a simple solution emerging from such study is remote.

All-embracing solutions to animal welfare problems are also unlikely to be produced by the remaining group of contributors to the debate, the biologists. Biologists investigate the problems of describing and measuring animal welfare in the hope of identifying verifiable biological relationships that can subsequently be put to use in improving animal welfare. However, even knowledge of these relations can provide no final answer to welfare problems, because of the ever-present moral component. When biological relationships are verified, each member of the public still has to decide on a moral position with respect to animals. The contribution of the biologist is to make it easier to identify sound biological positions on which one might take a moral stand.

None of these groups of people alone will solve problems about animal welfare. They will need to collaborate, to discuss the issues and subsequently take action to inform the public and legislators. All must be willing to reconsider their opinions in the light of new information and new arguments. The process of greatest importance in solving welfare problems will be education, both of oneself and of the community. Philosophers and biologists must analyse their disciplines in relation to animal welfare so as to understand better the nature of the problems, especially those aspects extending across disciplines. More than either group seems generally to realize, philosophical propositions can be reduced to nonsense by biological misunderstandings, and biological pronouncements wrecked by logical inconsistencies.

Integration of the disciplines is essential. The results of dialogue must percolate into the community and become part of public education. When political advisors can be given reliable information on which philosophers and biologists largely agree, there is a chance that governmental, public, commercial and private actions concerning animals will be widely approved by society.

The most conspicuous contributors to the animal welfare debate, the activists and lobbyists, will be evident as long as there is a need for them. They will continue to function as barometers of welfare problems. Critics will be muted only when there are no further grounds for criticism and society has accepted the prevailing level of animal welfare.

1.4 THE IMPORTANCE OF BEGINNING FROM BASICS

The more complex a problem, the more important it is to have a sound basis on which to build a solution. Philosophers in the animal welfare debate have, to their credit, chosen their various bases, such as advocating 'equal consideration of ... the interests of all beings with the capacity for suffering' (Singer, 1990), or proposing that 'any animal ... has a right to life' (Rollin, 1981). When argued through with reference to real life, neither of these basic propositions is readily or fully acceptable to the average person. However, from such beginnings one would hope to be able to follow logically through to conclusions, identifying the steps with which one does or does not agree. Given a clear exposition of an argument, a person should be able to decide how many of the sequence of steps they accept.

The biological implications of animal welfare are not yet so well understood; it is therefore even more important to establish basic concepts. A great deal of information is available about the physiological functioning of animals and a certain amount of knowledge is available about their psychology and behaviour. Measures of injury, pathology and immunological defences are relatively precise, and data on all of these can be used to help understand and assess animal welfare. A single notion that strongly conveys the current focus of biological interest in animal welfare is 'stress'. Stress in common parlance conveys the idea of an excessive physical or mental burden on an animal. Stress can overwhelm animals and cause suffering. Stress in animals is seen as a factor to be minimized as a matter of course in all animal management or other human–animal interactions. But is it so simple?

How well is the term stress understood? Can anyone say with confidence exactly when an animal, or even they themselves, are stressed? Consider a situation in which human or animal athletes run to the limit of their capacity. Are they stressed? Perhaps so, but if that effort improves a subsequent performance, was the first exhausting run 'stressful', or was it 'training', or was it both of these? Consider the case of an animal that is so protected from the physical and social challenges of the environment that subsequent exposure to the natural world causes it to collapse. Has the animal benefited, or suffered, from the lack of environmental buffeting? Has the isolation itself been a stress? A usable concept of stress must take account of such questions.

Analysing what biologists mean by stress is a logical place to begin a study of animal welfare. The task undertaken in this book is to define and relate stress and welfare, and to review ways of measuring and using them. Sufficient attention will be given to non-biologist's views of stress and welfare to ensure that the gap between biology and other disciplines is narrowed rather than widened.

First, two brief explanations of our approach are necessary. The term stress is in current usage in physics, physiology, psychology and pathology – with somewhat different meanings in each field. Unfortunately these various meanings have confused the use of the word in discussions of animal welfare. In Chapter 4, after development of basic theories in

Chapters 2 and 3, a proposal is made as to how the disparate uses of the term 'stress' might be reconciled. Until that point, it will be avoided when possible.

The second explanation concerns the extent of coverage of this analysis. Since all animals including humans are exposed to disturbing situations at some time, both human and other animal species will be considered. Plants can also be subjected to stress, but that will not concern us; use of the term 'welfare' is in practice limited to animals. An ideal model of stress and animal welfare would encompass all activities of all animal species. It would include animals ranging in size from protozoa to whales, and in activity from barnacles to cheetahs. We still have to consider the welfare of animals whatever their lifestyle. Some animals are predators, some prey, and some parasites. From a human viewpoint, some are pests, others are dangerous. They can be sources of companionship, food, clothing, work and entertainment. Some are unavoidable casualties under our feet. Others are displaced by our farms and our buildings. Some cohabit unseen in our houses. What is written here about non-human animals could be rewritten replacing the human perspective with that of each other species in turn, for all species will be treated as essentially equally deserving of our attention. So the range of animals to be covered is vast, and the range of interactions almost infinite. A simple unifying concept of animal welfare, which takes account of the abilities of the animals, is the only possible way of encompassing such dimensions. In order to produce this concept, welfare must be considered a characteristic of the individual, not as something given by one individual to another.

1.5 THE CHALLENGE AHEAD

The groups pondering the complexities of animal welfare have frequently pleaded with society for greater awareness of animals' needs, for further study of the questions raised by the animal welfare debate, and for widespread education based on the results of such study. The response of biologists must be to establish basic terminology and build logical arguments on it. In this book, the key terms stress and welfare are clearly defined within a logical and scientific framework, in order to facilitate the use of these terms. From this base, practical guidelines can be proposed for recognizing, measuring, avoiding and alleviating stress and its effects. The ethical question of what is a justifiable imposition on an animal will finally be considered, and a guide given on how to answer such a question.

Chapter 2
Systems regulating body and brain

2.1 THE BASIC CONCEPT OF HOMEOSTATIC CONTROL

Life within a cell, tissue or body depends on a supply of essential nutrients, mechanisms to process them, and arrangements for disposal of waste. Just as importantly it requires a network of control systems which adjust the inputs and outputs to ensure that chemical transformations occur in an environment of relative constancy. The vitality of all animals, from single-celled protozoans to multi-celled whales, depends ultimately on the efficient operation of the regulatory systems that control the conditions within their bodies. Homeostasis is an approximately constant state which varies only within tolerable limits; the regulatory systems which maintain the internal homeostasis of the body are said to be **homeostatic**.

If homeostatic mechanisms within a living animal were to operate with wheels and cogs, they would whirr continuously. At every instant of an animal's life, when it changes position, when the light or temperature levels alter, or when food becomes available, adjustment of function is required so that the state of the body remains stable. Stability of brain functioning is just as important as that of other parts of the body. Most alterations in the environment and the stimuli which result from them elicit appropriate responses, and homeostasis is maintained. When, by using behavioural or physiological adjustments or both, an animal is able to cope with environmental conditions, it is said to have **adapted** (see definition in Glossary). But sometimes adaptation to stimuli from the environment is not immediately possible because they are excessive in intensity or duration, or noxious, or novel and hence no system yet exists for coping with them. Inadequacy of the regulatory systems then leads to a displacement of state outside the tolerable range. The animal's attempts to adapt to its environment are inadequate, and injury, poor health, suffering or reduced survival chances will result.

At its most fundamental level, a study of stress and welfare is concerned with the stimuli that pose such challenges, how animals deal with such problems, and whether, as a result, an animal's efficient functioning and future are put at risk.

2.2 INPUTS TO CONTROL SYSTEMS

Convenient analyses of control systems are made by considering the various kinds of information passing into the system (**inputs**), the resultant responses arising from the system (**outputs**), and the processes that relate inputs to outputs.

2.2.1 Simple inputs

The most numerous inputs to an animal's regulatory systems come via its sensory nerves. The light impinging on an animal, the temperature of its surroundings, and pressures on its surface, together with sound, smell, taste and other less obvious factors such as gravity, apprise an animal of its environment via neural sensory mechanisms. Animals, other than the simplest, collect both non-specific information about the environment for example via proprioceptor (see Glossary) inputs which tell the individual about the relative positions of its limbs, and specialized information, such as sound, via the cochlea and visual patterns from the retina. Even simple organisms have sensory systems. These may respond to general environmental changes, or be specific, such as the sites receptive to particular molecules on the surface of the bacterium *Escherischia coli* (Morimoto and Koshland, 1991). Small changes in inputs from sensors can have important influences on an animal's activity, such as when the presence of specific molecules alters the tumbling sequences of *E. coli*, or light intensity influences the laying patterns of birds.

2.2.2 Complex inputs

Although some regulatory processes in the body are affected by simple inputs, most involve responses to multiple signals. Even initially simple inputs may be interpreted in the light of previous experience and compared with other inputs so as to build up a more accurate, and more complex, picture of the environment and of the relevance of changes in that environment for the individual.

In addition to providing information about the environment outside an animal's body, or the physical state of its body, some inputs to the system in the brain that controls behaviour and other interactions with the environment arise from another part of the brain, the emotional system. The significance of fear, anxiety and pain in this respect is so important that it is considered separately in Section 2.7. At this stage of our discussion of control systems it is important to emphasize that a single simple sensory change, such as the sight of a potential predator or rival, can be processed so that the ultimate input is very complex. It could initiate not only complex responses dependent upon previous experience, but new bodily changes such as the release of adrenal hormones which will themselves have an important effect on control systems.

2.2.3 Adaptation, habituation and sensitization

A characteristic of nervous mechanisms as they sense the environment is the decline in their response when they are exposed to repeated stimuli (Guyton, 1991). This exemplifies a second meaning of 'adaptation' (see Glossary): the waning of a physiological response, at the cell or organ level, to a particular condition, including the decline over time in the rate of firing of a nerve cell. A consequence of this adaptation of receptors is that an initial major imposition on an animal may be quickly reduced. This natural process allows an animal to differentiate certain types of stimulus, such as touch and vibration, and to avoid sensory overload which otherwise would make life unbearable. Sensors adapt at different rates. For example, vibration sensors adapt in fractions of a second, touch sensors within a few seconds, temperature sensors within minutes, and blood pressure sensors over a number of days. Pain sensors adapt even more slowly, and usually incompletely. The term 'adaptation' is also used in a third sense, in evolutionary biology, to describe any structure, physiological process, or behavioural feature that makes an organism better able to survive and reproduce than other members of the same species. It can also mean the evolutionary process leading to the formation of such a trait.

One of the processes which can help an individual to adapt, in the sense that it can cope with environmental conditions, is the learning process **habituation**. This involves the waning of an individual's response, which could still be shown, to a constant or repeated stimulus. Such a waning of response may be a consequence of a simple gating process reducing the efficacy of synaptic transmission in the nervous system. It may also be very specific to the stimulus concerned and involve the comparison of inputs with some neural model of previous inputs (Sokolov, 1960; Horn, 1967; Broom, 1968; Brown, 1991). An individual may become unaware of a sensory input despite its continuation. For example, people become unaware of the presence of clothing which initially stimulated many touch receptors. The sensory input is still there however, for a small change in input can lead to a response. Even quite elaborate responses may become habituated. For example, domestic animals may show a substantial initial reaction following quite gentle handling but show no such response after several such experiences (Beilharz, 1985).

Habituation occurs more readily to some types of stimuli than to others. The sound of a branch falling from a tree might elicit a startle response from an animal in a wood the first time it is heard, but frequent recurrence of the stimulus with no adverse effects would result in habituation. However, the same animal would be unlikely to habituate to the sight of a hunting leopard. Similarly, birds feeding in fields may not readily habituate to a sight or sound similar to that of a potential predator. Some bird scarers are probably effective because the bird experiences some uncertainty about whether the object perceived is dangerous (Fazlul Haque and Broom, 1985).

If a disturbing environmental event is detected, such as the sight of a hunting leopard, the responses to repetition of this may become greater rather than smaller. Clark (1960) showed that the response of the ragworm *Nereis* to repeated passing shadows habituated, but Evans (1965) demonstrated that there was an increase in the ragworm's response to successive presentation of shadows if it was also subjected to electric shocks which were not associated in time with the shadow presentation. Such an increase in response to continuing or repeated stimulation is called **sensitization** (Evans, 1965; Groves and Thompson, 1970; Brown, 1991) and may be thought of as a facilitation of the response (Guyton, 1991). This too is one of the normal learning processes in the central nervous system, one which will tend to exaggerate responses. Another example of sensitization is the increase in response to the light touch of a biting fly on the skin when this touch is repeated. The effect may also be seen when a dog, presented several times to a veterinarian for an injection, reacts more and more adversely, even though the stimulus situation is essentially constant.

Organisms can never fully adapt to some sensations, such as certain types of pain (Guyton, 1991). When they, or other sensations, remain incompletely adapted as continuing inputs to physiological control systems, it may be that the animal is considered unable to cope fully with its environment. With inputs continuing to the control system, outputs or responses from the animal will also continue to be evoked. The animal's response to continuous stimulation, especially if the stimuli are noxious or debilitating, will be examined further in Section 2.4.

2.3 MOTIVATIONAL STATE: STIMULATION IN RELATION TO LEARNING

The inputs which an animal receives as a consequence of changes in its world are interpreted in relation to previous experience. In other words, some event in the past has changed the brain in such a way that when a particular kind of sensory input is received, the information which it provides for the individual is different from that which such an input provided the first time that it was experienced.

There are various kinds of learning, some of which are discussed further in section 2.8. Habituation is one kind which has been discussed in order to explain how it is related to adaptation. Some habituation involves brain processes which must be just as complex as those involved in learning elaborate associations (see p. 39). The key point is that stimulation is very often linked with learning in some way. Most kinds of sensory input have been experienced before, so the one or many previous experiences will alter the input which actually reaches the centre in the brain where decisions are taken. The integration of inputs and the taking of decisions will be considered below.

The brain of an animal receives, analyses and interprets inputs so that the animal has a perception of its environment which can be used when taking decisions about what action to carry out next. Each input that provides information to the decision-making mechanism of the brain about an aspect of the animal's world can be considered a **causal factor** (McFarland, 1971), for it may alter the animal's future behaviour and physiology. The levels of some causal factors provide information about the world outside the animal, for example the brightness of light detected by the eye. Other causal factors reflect variables within the animal, such as blood glucose levels, the concentration of a steroid hormone in the blood, or the output from some internal time-clock. Each sensory signal or result of internal body monitoring is interpreted in the light of previous experience (Broom, 1981; Fraser and Broom, 1990). Hence causal factors are defined as inputs to the decision-making system, each of which is an interpretation of an external variable or an internal state of the body. The internal state referred to includes that of systems in the brain as well as that of other body systems. The causal factors which emanate from within the brain include 'ideas' which are not an immediate consequence of input from sensors, for example in the case of a dog which stands up and goes directly to the site of a buried bone and unearths it. There are also causal factors which are the consequences of more general brain activity resulting from many kinds of input.

If a dog is deprived of water, after some time there will be input to the brain from: (a) monitors of body fluid composition; (b) sensory receptors indicating a dry mouth; (c) probably from oscillators which have a regular output and might initiate drinking at particular times of day or after particular intervals; and (d) other brain centres which could make the animal aware that drinking has not been possible for some time. The change in the state of the animal arising from this group of causal factors is shown in Fig. 2.1. As the levels of these causal factors rise there will also be increases in the likelihood of drinking if the opportunity arises and in the extent of related activities which would promote water acquisition.

The actions of an animal will depend upon the levels of many different causal factors. This can be represented in a diagram (Fig. 2.2) which depicts for a dog the interaction of two sets of causal factors – those resulting from water deficit and food deficit – in what is known as 'causal factor space' (McFarland, 1971; Sibly and McFarland, 1974).

The decision of an animal about what to do will depend upon its position in this causal factor space with respect to the two axes, that is, to the two sets of causal factors. In Fig. 2.2 a dog whose state has reached *B* is more likely to eat and to work in some way to get food than one whose state is at *O*, while an animal whose state is at *A* is more likely to

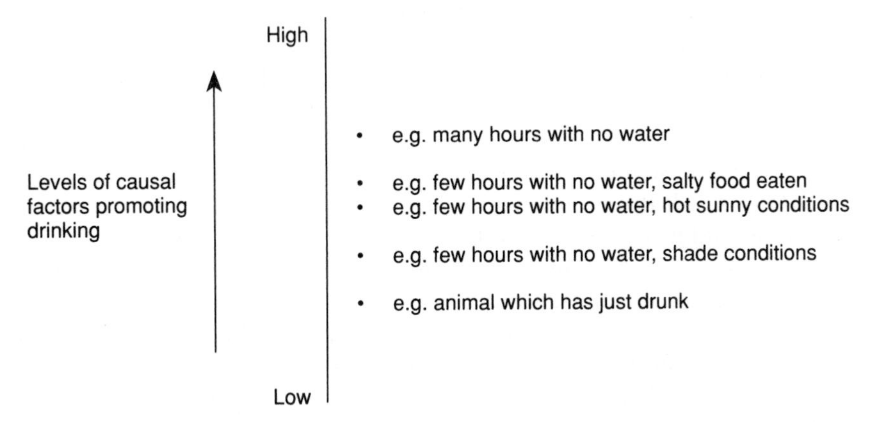

Figure 2.1 Levels of causal factors which promote a particular action vary over a range, and the state of the animal can be described in terms of these (after Fraser and Broom, 1990).

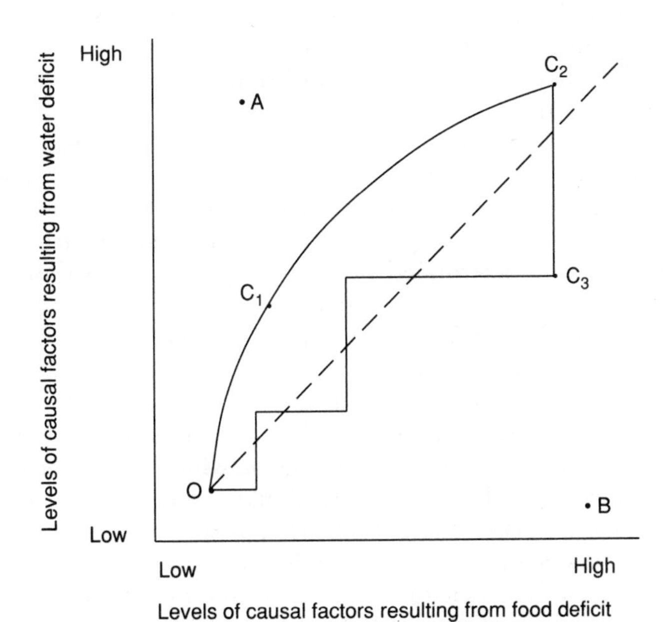

Figure 2.2 Motivational state of animals A, B, and C in two-dimensional causal factor space. Animal A is most likely to drink, whereas animal B is most likely to eat. The changes in state of animal C are explained in the text (after Fraser and Broom, 1990).

drink than one whose state is at O. Plotting the state of the animal in this two-dimensional space allows interactions between the two sets of causal factors to become clear. When a dog is deprived of water its state moves up towards A on the state space plot but it also moves to the right because dogs given no water cannot eat as much. The change in state of the animal as a consequence of water deprivation is shown as a trajectory from O to C_1. If this animal were then deprived of food as well as water, its state would move sharply to the right and up further to C_2.

The behaviour of an animal which is given the opportunity to either eat or drink will depend upon its state as represented in Fig. 2.2. A dog whose state is at C_2 is a little higher on the water deficit side so it might drink. This would bring its state down across the boundary line to C_3 at which time it might switch to eating, thus lowering the causal factors resulting from food deprivation. An example of a possible course of the animal back to O is shown. The actual paths chosen by animals whose state was manipulated in this way have been described and the nature of the decision-making mechanism explored in papers by Sibly (1975), McFarland and Sibly (1975) and Sibly and McCleery (1976). The position of the boundary line for switching from feeding to drinking could be altered by making the animal search harder for the food or use more energy to get the water (Larkin and McFarland, 1978). In reality an animal's decisions will depend also on many other causal factors, each with its own dimension. The **motivational state** of an individual is a combination of the levels of all causal factors in the brain, or more technically, its position in multi-dimensional causal factor space.

An alternative explanation of this idea for people unfamiliar with the terminology might be: animals are constantly under pressure to embark on various activities (e.g. drinking, grooming, mating, etc); which of these is undertaken will be determined by the strengths of the various sorts of biological and social pressure; the activity or activities finally pursued would be that, or those, for which the overall pressure is greatest at the time. The dog referred to in Fig. 2.2 might have recently eaten salty food and hence there would be causal factors resulting from the taste of the food, the concentration of blood or saliva, the delay since drinking last occurred and the visual cues from a water source, all of which would promote the likelihood of going to drink. However, drinking might be delayed because of other causal factors resulting from the approach of a person carrying a stick, or the odour of a bitch, or the detection of a very palatable food item. All of these causal factors are a part of the motivational state of the dog, and one or more of them will determine which action is taken next.

In summary all actions except the simplest reflexes are determined by an animal's motivational state. An understanding of motivational processes is therefore fundamental to the assessment of animal welfare using behavioural and physiological measures. Further information on this topic can be obtained from Broom (1981), Fraser and Broom (1990), McFarland (1985), Toates (1986) and Colgan (1989).

2.4 OUTPUTS FROM DECISION CENTRES

An understanding of the cause and nature of output responses from regulatory systems after the motivational state has been altered is essential for the study of stress and animal welfare. Such responses are not only indicators of the adjustment of an animal to its environment, but also potential indicators of maladjustment.

2.4.1 Neural and muscular outputs

In more complex animals, outputs from decision centres commonly appear as various forms of neural and muscular activity. These may be no more than an electrical spike in the brain waves – shown on an electro-encephalogram (EEG) – as the signal reaches another system in the brain, and they may cause no other physical sign. Such a sensory perception committed to memory may scarcely be measurable, yet it may exert a profound influence on the shaping of future responses.

Most of what we call responses, however, are changes in whole body position or pattern of locomotion brought about by nervous control of muscular movement. A slight change in posture or a jump to catch prey might be the responses to sensory signals from aching joints or the sight of a darting insect respectively. Less complex animals have proportionally less machinery to handle outputs, though even the simplest have mechanisms to co-ordinate their physical activity and retain information about past events. Again for simplicity we shall elaborate only on the output mechanisms of the more complex animals, assuming that many of the principles will extend to even the simplest.

The neuromuscular responses in some cases are structurally coupled to input signals that follow invariably upon a particular stimulus. These are the reflex responses. They are unquestionably of survival value in eliciting an instant response to a noxious stimulus, but the unvarying nature of the response introduces an inflexibility which can reduce an animal's capacity to adapt.

Certain animal welfare problems may manifest themselves as integrated nerve and muscle outputs, the measurement of which may serve to quantify the severity of the problem. For example, repetitive behavioural routines, such as stereotyped bar-biting by individually stalled or tethered sows are co-ordinated muscle movements which may well arise from maladaptation to the environment or malfunction of the sensory, integrative, decision-making or motor system. In all of those, except the motor malfunction, we need to know about the motivational state in order to understand the behavioural abnormality. What can be changed in the animals or the environment to break this cycle of apparently inappropriate behaviour? These are the types of question one finds at the core of many animal welfare problems.

2.4.2 Hormonal and neurohormonal outputs

The consequences of certain environmental events which should be activating a homeostatic control system may not be outwardly evident to a human observer. Many occasions arise in which a hormonal response is evoked but there is no immediate evidence of a change. As examples, eating a large meal induces insulin release from the pancreas, and decreasing air temperature triggers the release of metabolic hormones from the thyroid gland. Externally the effects are not visible for some time, if ever, yet the mechanisms contribute significantly to an animal's adaptation. Because they can be measured, such hormonal changes can provide insights into the processes of homeostatic control and physiological adaptation.

A response of particular significance in the adaptation of more complex animals is that described by Cannon as the 'fight or flight' reaction (Cannon, 1935; Guyton, 1991). This involves both neural and hormonal outputs. Excitement, anxiety or alarm, initiated either within the animal or from an external challenge, result in a co-ordinated response of the autonomic nervous system including the medullae of the adrenal glands. The autonomic nerves together with the adrenal hormones adrenaline and noradrenaline (called epinephrine and norepinephrine in North America) induce acceleration of heart rate and blood flow and a set of metabolic adjustments which prepare the animal physiologically to cope with an emergency. This alarm reaction, while unquestionably a beneficial and even crucial output from control systems in many situations, is relatively inflexible and, like some nervous reflexes, may commit an animal to a response even when it is inappropriate.

The adrenal cortex also has a central role in the hormonal response to disturbance of homeostasis. Following various types of both internal and external stimuli, these outer segments of the adrenal glands will release glucocorticoid hormones affecting energy and protein metabolism and immunological reactions, or in other circumstances, mineralocorticoid hormones affecting body fluid balance. The first stage of the response of the hypothalamic-pituitary-adrenal cortex axis response is the activation of corticotrophin releasing factor (CRF) production in the hypothalamus by interleukin lß. This results in the release of adrenocorticotrophic hormone (ACTH) from the adenohypophysis (anterior pituitary). ACTH is transported in the blood to the adrenal cortex where cortisol and corticosterone production occurs. In some animals, only one of these glucocorticoids is produced and in others both are produced.

2.4.3 Integration of experiences over time

The consequence of an animal receiving a sensory signal about its environment may be a reaction as simple as the flicker of an eyelid, or as complex as an emergency response under the influence of catecholamines comprising physiological adjustments of the cardiovascular, respiratory, metabolic and hormonal systems, together with some integrated behavioural response. In the short term these responses usually serve the function of promoting homeostasis in the animal.

Over longer periods such responses have further significance for any animal with a memory, and that probably includes all animals. Actions with consequences that are pleasant, neutral, or unpleasant may be remembered, and any subsequent response altered as a result. The commitment to memory of a sensory experience may not be immediately evident, but the effect on ensuing responses can be profound. The role of past experience in shaping responses of the regulatory system is of great importance in determining the impact of fear, loneliness, anxiety, aggression, commitment, satisfaction and so on, which have significant influences on animal welfare. More will be said about the influence of past experience in Section 2.6.

2.5 CONTROL SYSTEMS AND NEEDS

2.5.1 Simple models of control

Control systems are mechanisms which functionally connect inputs to outputs. In mammals some operate via simple nervous linkages, others via complex nervous linkages through the brain, involving hormonal effects in many cases. Those which operate at more basic levels will not concern us here. Even working with nerves and hormones introduces complications for, in practice, the operation of nerve networks is influenced by hormones, while hormone secretion is simultaneously affected by nerve activity. Almost invariably control systems involve both nervous and hormonal influences.

Simple controls operate within nerve fibres themselves, in that they either fire and carry an impulse, or they do not. A stimulus that does not bring a nerve to its firing threshold disappears, and is filtered out. A somewhat more complex control is exemplified by the reflexes that constantly modify input–output relations of skeletal muscle via neural links in the spinal cord and central nervous system (Guyton, 1991). Contraction of skeletal muscle is induced by impulses arriving in motor neurons. But these are modified by signals reflecting the degree of muscle stretch sensed by the spindle receptors in the muscle, and by Golgi tendon organs which inhibit contractions that approach the mechanical limit of the muscular system. Another complexity is that muscle actions are modified by the composition of the body fluids, for example by the concentration of thyroid hormones (Guyton, 1991).

The complications of this control are multiplied many times over when body functions depend on the processing of signals into and out of the higher nervous system. In such cases, numerous inputs, both facilitatory and inhibitory, converge on the controlling centres to determine the output. Furthermore, the output is influenced by the composition of the surrounding fluid, including its content of any cyclically changing sex hormones, as well as by previous experience through the influence of memory.

Hormonal control systems can also operate in a simple fashion. Insulin, which reduces blood glucose concentration, is produced from the pancreas in proportion to rises in glucose concentration, and hence prevents large rises in blood glucose from occurring, for example after eating. However, this control, though predominantly hormonal, is also subject to influence by the sympathetic nervous system (West, 1990). Thus hormonal control systems are subject to nervous influences, just as nervous control systems are affected by hormones. These two components of the response of an animal to its environment can be viewed as elements of a combined control system that determines the animal's physiology and behaviour.

Simplified physical models of the relations between inputs and outputs can be depicted as in Fig. 2.3. Other systems relating inputs to outputs also occur, for example so-called 'on-off' controllers (Bligh, 1973), but the type shown in Fig. 2.3b with a proportional controller and adjustable set point is one that seems to occur widely. It will be used later as a model on which to base discussions of stress.

The simplest proportional controller (Fig. 2.3a) can be drawn as a rigid rod pivoting on a fulcrum, linking inputs to outputs in direct proportion. A more refined version of this system (Fig. 2.3b) operates in the temperature controller of a domestic central heating system. The air temperature is sensed at various points in the house, causing heating to be adjusted so that the colder the house the more the heaters are activated (this is negative feedback control see p. 25). The temperature signals entering the system are integrated, and collectively they control a heat output which varies in magnitude or duration so that it is **proportional** to the input error.

The magnitude of the response for a given change in input is called the **gain**. In Fig 2.3a, the gain depends on where the fulcrum is placed under the lever. The gain may vary from one control system to another, and may change from time to time within one control system. The gain is therefore adjustable. Since the fulcrum under the rod linking input and output in Fig. 2.3a is closer to the input end than the output end, small changes in inputs bring about large responses; in other words this system as depicted has a high gain.

The output, or response, in such a system is not simply turned on or off. It operates at an intensity dependent on the strength of the input signals. To achieve this, there must be within the regulator system (Fig. 2.3b),

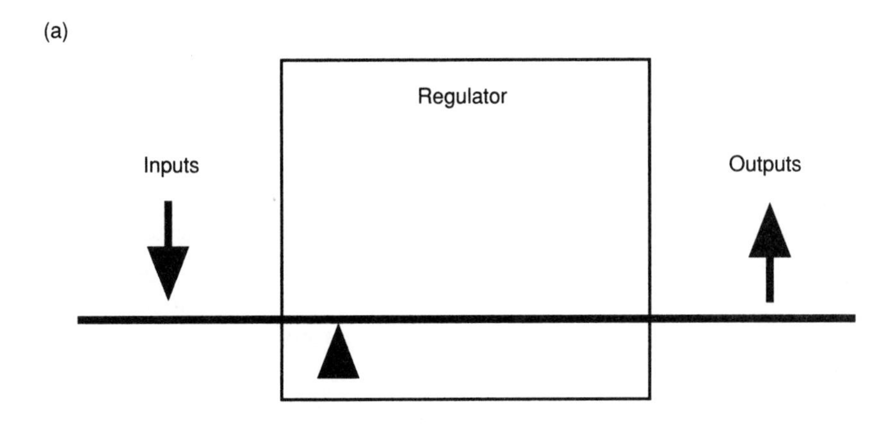

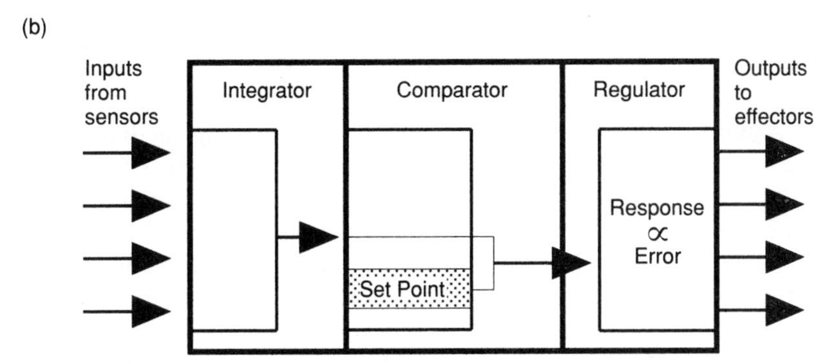

Figure 2.3 Models of biological control systems. (a) rod pivoting on fulcrum (b) also a proportional controller but like that of a domestic central heating system in that it has an adjustable set point and gain (after Bligh, 1973).

components that integrate the inputs, compare them to a set-point or optimal value, and co-ordinate the outputs.

The output to effectors can change either because the input has changed, or because the set-point for control has altered. This last change is akin to turning the thermostat of a central heating system up or down. Many physiological mechanisms have the capacity to change as though the set-point was being altered. A control system of this type is described as being a 'proportional controller with adjustable set-point and gain'. Examples will emerge in later discussions of an animal's responses to incoming stimuli being varied by adjustments which appear to be due to alterations of the set-point or of the gain of the control system.

2.5.2 Motivational state as the determinant of action

The notion that an animal acts to stabilize its internal environment via an assortment of homeostatic systems was introduced in Section 2.1. The proposal was developed in Section 2.3 by reasoning that an animal achieves this by integrating a myriad of inputs, including those from sensors, which may be as simple as those resulting from changes in light intensity, or as complex as the recognition, following experience, that a certain site in a field will provide shelter. Inputs from external sensors, internal monitors or rhythm generators and from other brain centres are the causal factors which together make up the motivational state. It is this state which leads to an animal carrying out particular activities. Furthermore, changes in causal factors may modify not only the motivational state but also the set-point and the gain of an animal's regulatory systems.

A hypothetical example may be the best means of illustrating this concept. Consider a sheep at pasture in hot summer conditions which has not drunk for 12 hours. Its body fluids are becoming concentrated and several causal factors are becoming stronger and increasing the likelihood that the animal will drink. These may include inputs from monitors of body fluid concentration, mouth receptors, the memory of the time since the last drink, and temperature receptors. In other words, the sheep is becoming more and more thirsty. The more concentrated the body fluids, the stronger the animal's interest in seeking water. If water were near, the animal would drink. But as other sheep in the flock have not yet moved to the water trough, the animal is disinclined to walk there alone. The normal motivational output from the body fluid regulatory system is outweighed by the causal factor resulting from the animal's preference to stay with the flock. A body fluid concentration greater than that which usually occurs at this time of day is required to prompt this animal to drink. Gregariousness has raised the threshold that must be reached before the animal will drink. In control terminology, it is as if the set-point has been raised. Finally, other sheep begin to walk to the water and the constraint on the focal animal not to move alone is removed. The set-point returns to normal.

At this stage, the same sheep has higher levels of causal factors promoting drinking than do other animals in the flock, and it runs faster than they do. A further control process now determines how much it will drink. Since the weather is warm the resulting causal factors have some influence. The warmth indicates, perhaps without any higher level concepts being involved, that a greater water input than usual may be required. The animal thereby assesses its future needs and drinks more water than usual for a given elevation of body fluid concentration. The drinking response is enhanced or, put another way, the gain of the control system is increased. This ability to change the gain so as to prepare for probable future needs is a form of feedforward control (see Section 2.6.2).

This example, despite its complexity, is a simplification of the operation of the type of system believed to control animal behaviour. Nonetheless, it serves to illustrate the enormous variety of inputs, and how events might be interpreted in terms of alterations in stimulus input, set-point or gain. Every aspect of an animal's life is under a control system as complicated as this, including simple responses such as standing, lying, scratching and chewing. Complex activities such as foraging, predation, mating or aggression must require even more complicated integration and processing.

2.5.3 The concept of needs

Animals have functional systems controlling, for example, body temperature, nutritional state and social interactions (Broom, 1981b; Fraser and Broom, 1990), and there will always be inputs to these systems. By further investigating functional systems and motivational mechanisms we can go some way towards identifying the resources or stimuli in the environment that are required by or important to animals, and so learn something about an animal's needs.

When do we say that an animal needs something? To **need** is to have a deficiency, often manifested as a homeostatic maladjustment. A need can therefore be defined as a requirement, which is fundamental in the biology of an animal, to obtain a particular resource or respond to a particular environmental or bodily stimulus. Hence, some needs are for food, water or heat, but others are for a certain behaviour such as grooming, exercising or nest building to occur. Control systems in animals seem to have evolved in such a way that, in some circumstances, animals have a requirement to perform certain behaviours and are seriously affected if unable to carry them out.

At any moment an individual will have a variety of needs, some of greater urgency than others. Each is a consequence of the particular motivational mechanisms of the individual (Baxter, 1988; Hughes and Duncan, 1988a,b; Broom, 1988a). Some needs are simple, such as to escape the debilitating effects of high concentrations of body fluids or a high body temperature. Others are complex consequences of the mechanisms which promote survival and reproduction, for example, the need to circumvent the deficiencies in mental functioning which result from too little variety in sensory input or insufficient contact with other members of the species (Broom, 1991c, 1992).

Some reports and legislation refer separately to physiological needs and to behavioural or ethological needs. Needs may be recognizable because of effects on the physiology or behaviour of animals, and animals may have a need to show a certain behaviour (Toates and Jensen, 1991), but the need is a requirement, usually to remedy a deficiency as explained above, and is not itself physiological or behavioural. Hence it is scientifically more precise not to qualify the term 'need' except with the word 'biological'. It may be desirable in legislation and in advisory codes to place some emphasis on the existence of many needs which can only be satisfied if the individual is able to perform a particular behaviour. This can be done by defining biological needs as above, or by referring to 'biological needs including those satisfied only by showing specific behaviours'.

When an animal has an unsatisfied need, its motivational state will usually elicit behavioural and physiological responses that remedy that need, so the individual will be able to cope with its environment. If a need cannot be satisfied, the consequence in either the short term or the long term will be poor welfare. Indications of the nature of the needs of a particular animal are often deduced from situations where there is some inadequacy in the environment. In fact, this is a way of investigating what constitute needs. Another way of finding out about the needs of animals is to examine what they do when they have some free choice of environment (Chapter 7; Fraser and Broom, 1990).

A problem associated with the use of the word 'need', especially in legislation, is that the deficiencies involved range from the rapidly life-threatening to those which are relatively harmless in the short term. The problem is more clear-cut in German where the word *Bedarf* means a need which must be satisfied if life is to continue, whilst *Bedurfniss* means something which the individual wishes to be satisfied. This conceptual distinction is referred to by Dawkins (1990) in a discussion of needs and wants. If providing for the needs of an individual person or domestic animal were limited to the *Bedarf* meaning of needs, then the quality of life of the individual would be poor indeed. As discussed in Chapter 7, most of what is strongly avoided is harmful and most of what is strongly preferred is beneficial, so *Bedurfniss* will usually be equivalent to need as defined here. However, some of what is wished for is not necessary, in the sense of essential for life, or it is harmful, so the reference to 'fundamental to the biology of the animal' in our earlier definition of 'need', and to 'deficiency' in the definition of 'to need' is valuable. Whichever word is used in German legislation should be defined so as to take account of these problems.

2.5.4 Motivational dilemmas and the 'trade-off' concept

When the control systems affecting the various aspects of an animal's life are examined in isolation, the outcomes of their operation are commonly predictable. But in most natural circumstances, two or more systems interact, often driving different variables towards independent goals, and the outcome of the interaction is then far more difficult to predict. Sometimes control systems function in harmony to achieve a common purpose, as when increased blood circulation in the skin and increased respiratory ventilation jointly serve to keep an animal cool on a hot day. On other occasions, the outputs of two separate control systems conflict. This occurs when the achievement of homeostasis requires two responses that cannot be pursued simultaneously. For example, an animal cannot both forage for food and escape from cold if food is available only in the cold (Johnson and Cabanac, 1982). Other problems arise when one physiological apparatus services the needs associated with two physiological control systems. The cardiovascular system provides both skin vasodilation and muscle blood flow during exercise, and may be inadequate to provide both during exercise in hot weather. In such circumstances circulation will be compromised in both vascular beds (Johnson and Hales, 1984).

Extensive analyses have been undertaken of how animals cope when faced with behavioural dilemmas (McFarland, 1985). In simple terms, animals make choices which suggest that the costs and benefits of various behavioural options are evaluated, and certain options are 'traded off', that is partially or totally abandoned, in favour of the option which offers greatest gains. Humans face the situation if we need to go out to buy food in miserable weather. We weigh up psychologically how strongly we crave the food and compare that with the discomfort that must be experienced to gain it. A decision is made by evaluating the relevant causal factors including those resulting from assessing what alternative food is in the larder, how effective is our overcoat, how hungry we are, and so on.

Such dilemmas are also encountered by members of other species, sometimes with outcomes of relevance to animal welfare. In nature animals may be driven to tolerate unpleasant conditions while they pursue other goals. Newts, which court their mate underwater, face a dilemma during courtship between the needs to come to the surface to breathe and to continue a sexual encounter (Halliday and Sweatman, 1976). The extent of animals' voluntary tolerance of unpleasant stimuli in such situations of conflict is surprising. Rats will make excursions into a very cold environment (-15 °C) to eat tasty food even when nutritionally adequate food is continuously available in unlimited amounts in warm conditions (Cabanac and Johnson, 1983). We could interpret this as meaning that, in the pursuit of pleasure, animals will voluntarily tolerate considerable discomfort.

2.6 TYPES OF CONTROL

The impact of a disturbing stimulus on an animal is critically affected by how long the stimulus lasts, how often it occurs, and whether the animal was prepared for it in advance. Time is an important factor in the operation of control systems. A clap of thunder is a shock, not because it lasts for long, occurs often, or even solely because it is loud. It is alarming particularly because it is infrequent and unexpected. Yet the effect is brief and almost always harmless. A dog which is frightened by thunder may also be caused anxiety by the soft whirr of hair clippers if it is heard each time the animal attends a clinic for an injection. The associations are not made because it is a high intensity stimulus but because of its timing.

2.6.1 Rates of neural and hormonal response

The neural response to a peal of thunder peaks in a few seconds. Signals pass to and from the control systems along nerve pathways in a matter of milliseconds. The muscular startle response occurs in the same instant as, or even before, human subjects report perception of the sound. So too might the outputs elicited in response to brief noxious stimuli, such as butts from an aggressive littermate or shocks from an electric stock prod.

But not all nervous reflexes are so mechanical and rapid. Most incoming stimuli undergo comparison with past experiences before being transformed into complex behavioural responses, and may occur with considerably greater delay than the response to thunder. The co-ordinated response may be effected by the sympathetic rather than the somatic nervous system, as when the heart rate is raised, for example. In such a situation, the effect will last five to ten times longer, and additional tissues may be stimulated through the release of hormones from the adrenal medulla.

All unpleasant stimuli, even if brief, are likely to elicit some hormonal response. Not only will the 'alarm' reaction release adrenaline and noradrenaline from the adrenal medulla, it may lead to the release of glucocorticoid hormones from the adrenal cortex (Guyton, 1991). These hormones reinforce and extend the effects of the more rapid nervous reflexes. They do this partly by their slower release and especially by their slower turnover in the body. Adrenaline and noradrenaline take several minutes to disappear to half their initial concentration; glucocorticoids can take more than an hour. Consequences of noxious stimulation may thus last for hours. Other hormones such as prolactin (Krulich *et al.*, 1974), vasopressin (Anderson *et al.*, 1989; Guyton 1991), ß-endorphin (Guyton, 1991) and those from the thyroid (Guyton, 1991) may be mobilized, thereby further prolonging the effect (see Chapter 5).

2.6.2 Feedback and feedforward controls

Homeostatic mechanisms control the essential variables of the body, minimizing deviation of their levels from the optimal range. Where this is achieved by mechanisms which respond after changes occur in these values it is called **feedback**. If the response counteracts a disturbance to an individual, and can thus be considered to be opposite in sign to the displacement, it is negative feedback. If the response increases the displacement, this is positive feedback which is not homeostatic. In negative feedback the disturbance to the body occurs before a counteractive response begins. If the disturbance is very great or rapid in onset, substantial disruption may occur to the animal before the disorder is corrected (Fig. 2.4).

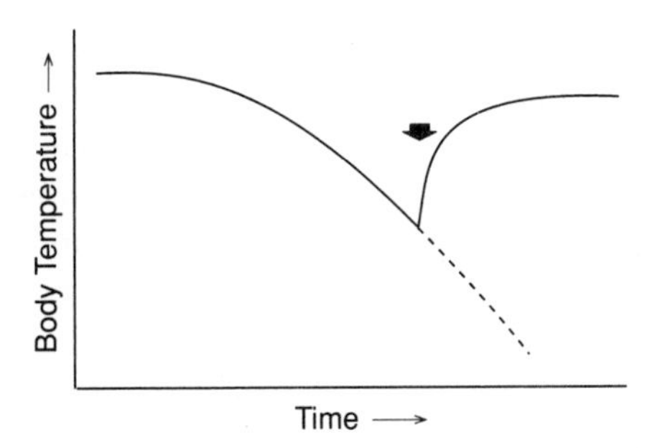

Figure 2.4 In negative feedback control a correction is made after the state of the animal has changed, and this restores the state to the former condition. Above, a drop in body temperature is detected, and corrective behavioural or physiological action is taken at the point marked by the arrow. The dotted line shows how the state would change if no correction occurred (after Fraser and Broom, 1990).

Feedforward controls also operate in many homeostatic systems, presumably due to their capacity to reduce such disruption. In these cases, an anticipatory mechanism induces a change in the homeostatic control system before a disruption actually occurs. The disturbance, when it comes, thus causes less upset to the animal (Fig. 2.5).

How does an animal anticipate a disturbance? One means is via a signal, indicative of a potential change, coming from a sensory input that is peripherally related. The sight of snow outside a window of a house will be sufficient to alert a person to put on a coat before venturing into the cold. It may even reduce the rate of cooling by limiting skin blood flow, before there is any lowering of body temperature. Other environmental challenges come on a regular daily or seasonal basis, so may be signalled by day–night changes or by changes in day length (Huntingford and Turner, 1987; Becker, 1987). By such means, an animal can prepare for the predators of the impending night, or the heat of the approaching summer. Any changes that constitute a threat to an animal's life will be particularly effective in alerting it to the possibility of that threat being repeated. A hunted species may be physiologically prepared

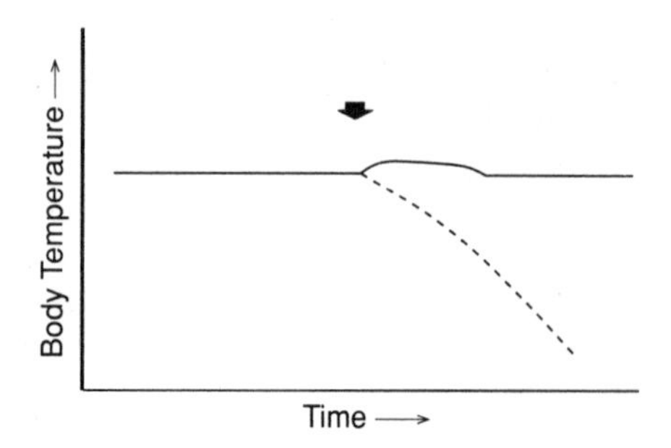

Figure 2.5 In feedforward control a change in state is predicted and corrective action taken before it can occur, so that the state changes little from its former condition. Above, a drop in body temperature is predicted and behavioural or physiological action is taken at the point marked by the arrow. The dotted line shows how the state would change if no correction occurred (after Fraser and Broom, 1990).

for flight by a distant gunshot or the sight of fleeing prey. When Metz (1975) studied food intake and inter-meal intervals in cattle given food *ad lib*, he found that the amount eaten in a meal was related more closely to the interval before the next meal than to the interval after the last meal. Similar results have been found in other species. The animals did not accumulate a nutritional deficit and then rectify it, but rather ate a meal in order to prepare for a period when they would not eat. Another example is drinking during or soon after a meal which occurs prior to the dehydrating effects of food in the gut which results in water being taken from the tissues by osmosis. There are many cases of animals behaving in a way that prepares them for future events rather than responding to previous occurrences. The effects are that many predictable changes in state are minimized or prevented.

Understanding the nature and causes of feedback and feedforward controls is important in studies of stress and animal welfare for two reasons. First the characteristics of negative feedback mechanisms determine the extent of disruption to an animal brought about by an environmental imposition. Inability to correct displacements of state from the tolerable range will usually be disturbing and will sometimes be damaging: we shall propose in Chapter 4 that an animal has a limited capacity to cope with such disruptions. Second, feedforward mechanisms, being anticipatory, tend to minimize the degree of disruption. Study of them should provide clues to the nature of successful adaptation. An inability to carry out feedforward control because of conditions imposed on an animal by man may be a source of particular frustration.

2.6.3 Predictability of stimulation

When an animal is exposed at constant intervals to stimuli that are not noxious or excessive, its response to them progressively wanes as it habituates (Section 2.2.3). Cattle grazing beside a road, although initially interrupted by passing vehicles, soon cease to show a response. However, if stimulation occurs irregularly, or with widely varying intensities, each stimulus continues to elicit a response. The progressive decrease in concern about a regular harmless stimulus may depend on feedforward control; the animal is prepared for the rumble of a passing vehicle and pays it no heed. But when the next stimulus is unpredictable either in time, intensity, or both, yet is certain to arrive sometime, the animal can prepare only by being constantly ready. The anticipation in such circumstances engenders a state of anxiety, and heightens the reaction when the stimulus is eventually perceived. In extreme circumstances an animal may even respond to its uncertain situation by showing learned helplessness (p. 56).

2.7 PAIN, FEAR AND ANXIETY

Pain is classified as one of the primary sensations which results directly from the activity of sensory receptor cells, either specialized for this function or for other functions. It deserves special consideration because there are important differences between it and other sensations, and it is particularly significant in animal welfare studies. Public concern about welfare has developed largely from a widely held conviction that animals experience pain much as humans do, and that pain is probably as distressing for other animals as it is for humans.

First we must consider what pain is. People asked this question firstly describe pain as something which they feel, as a sensation. There is always the qualification that the sensation is unpleasant. To recognize or measure unpleasantness we could examine the extent of withdrawal from, or avoidance of, unpleasant stimuli. But there are situations which are unpleasant without being painful, for example, those that produce fear. These situations involve more complex mental processes than just sensation. Fear follows relatively complex cognitive processes in which sensory inputs are related to previous experiences but pain itself is felt without such higher nervous activity.

Pain is a sensation which, without involving higher-level brain processing such as that associated with fear, is very aversive. Pain usually involves the specialized nociceptive neurons and often involves some degree of injury. Pain normally elicits protective motor and vegetative (basic bodily) reactions, causes emotional responses, results in learned avoidance behaviour, and may modify social and other behaviour. Detection and assessment of pain in animals relies heavily upon a combination of behavioural and physiological indices. On the basis of this definition, extremely bright lights, loud noises or some very strong smells could be considered painful. These inclusions are reasonable in relation to normal usage of the word.

Other definitions of pain which are quoted are often similar to that of Zimmermann (1985): 'Pain in animals is an aversive sensory experience caused by actual or potential injury that elicits protective motor and vegetative reactions, results in learned avoidance behaviour and may modify species specific behaviour, including social behaviour.' The problems with this definition are: (a) that it does not exclude fear; (b) that it refers to actual or potential injury when in reality there may be none, as in pain from a 'phantom' limb which has been amputated; and (c) that it refers to other consequences which may occur but need not do so. However, some of Zimmermann's phraseology is incorporated in the explanatory sentences following our definition above.

Can we offer a simple answer to the question, do animals feel pain? There can never be a conclusive answer, but currently no scientific information contradicts the points noted in the Brambell Report of 1965 that 'all mammals may be presumed to have the same nervous apparatus which in humans mediates pain. Animals suffer pain in the same way as humans.' Birds are of similar complexity to mammals, and fish have functional pain receptors (Rovainen and Yan, 1985) and similar physiological responses to painful stimulation to those shown by man (Verheijen and Buwalda, 1988). Many other animals have less complicated nervous systems and some species have no discernible nervous structures at all. However, even the simple protozoa have sensory capabilities and exhibit rudimentary behaviour (McFarland, 1985). Bacteria like *E. coli* (Morimoto and Koshland, 1991) withdraw quickly from noxious stimuli. Their immediate reaction, therefore, is like that of complex animals, and in the absence of evidence to the contrary, they must be assumed to have mechanisms for sensing hostile surroundings. Such sensitivity could imply, but does not prove, the existence of pain.

For simplicity we shall now leave the question of pain in simple animals and return to the discussion of pain in those animals with a nervous system. In such animals the existence of pain is believed to be indicated by a display of strong aversion on the part of an animal, which may be associated with fear. Pain may also be recognized and assessed by making measurements of nociceptor activity or by certain physiological changes.

The intensity of even a simple stimulus such as light can be increased to a level where it becomes irritating, then noxious, painful or harmful. At one time, this reaction was believed to be due to overstimulation of the various receptor types. Current evidence (Guyton, 1991; Janssens, Rogers and Schoen, 1988; Ganong, 1987) indicates that pain is mediated by specific receptors and nerves. Pain sensation is transmitted rapidly and discretely from certain specific receptors, and slowly and diffusely from other sites. It arises particularly where tissue is damaged, thus releasing degradative products. However, perception of causes of pain may be misleading since pain can appear to arise in an amputated limb. It is also possible for tissue damage to result in no pain. Humans have reported being unaware of pain during fighting despite having major injuries; likewise dogs injured in fighting can also appear oblivious of pain.

Pain sensation is transmitted along particular pathways in the nervous system in a similar way to sensations of temperature and touch, but it differs from them in the greater biological significance it conveys. The receptors, neuronal pathways and specific transmitter substances which convey information about pain are called the **nociceptive system**. It is apparent from the comments above that the link between the subjective experience of pain and the nociceptive input to the brain is not straightforward. This is due, in part, to the existence of endogenous opioids (morphine-like substances) in the brain which can have an analgesic effect. The sensation of pain can be suppressed, for example in cases of extreme injury, by the action of these opioids at particular receptor sites.

Emphasis so far on the noxious effects of pain may give the impression that pain should be prevented at all costs. That is far from the case. To suppress pain would in many situations be disastrous. The nociceptive system provides crucial information which in certain circumstances promotes recovery from injury and on other occasions initiates withdrawal and therefore aids survival when an animal encounters potentially lethal stimuli. When an animal has been injured, acceptable veterinary treatment will commonly leave the animal with some pain sensation for the express purpose of discouraging it from moving about, and thus promoting healing. Providing complete analgesia may permit an animal freedom to aggravate an injury.

A more basic biological effect of pain, revealed by all animals from protozoa to mammals, is that it causes them to withdraw from noxious, and usually harmful, stimuli. By doing so, they live to perpetuate their species. Thus the nociceptive system promotes survival. Animals need to experience pain, or at least aversive stimuli, in order to escape damaging conditions, to produce memories of pain that help them to evade such conditions in the future, and to survive to produce the next generation. Efficient operation of a nociceptive system is essential for survival.

Pain is complex in its effects and important in its impact on the entire range of animals. Some of the indicators of the intensity of pain are discussed in Chapter 6, where consideration of attempts to measure pain should further clarify the concept.

2.8 DEVELOPMENT OF REGULATORY SYSTEMS

2.8.1 Early preferences and abilities

Even at birth animals have an impressive repertoire of capabilities. One of the least developed of mammalian young, the new-born kangaroo, has the capacity to sense its environment and make its way through smoothed fur to its mother's pouch. More precocial species such as lambs, foals or domestic chicks attempt a range of behaviours within minutes of birth. Newly hatched domestic chicks can stand, walk, preen, peck, vocalize in several ways, approach visual or auditory stimuli and respond to contact with a brooding hen (Broom, 1981a, 1981b). Some of the early abilities and preferences of chicks develop in all individuals which are exposed to normal physical conditions during incubation. Others, such as responses to flashing lights or sounds, or to the voice of the mother, are affected by specific pre-hatching sensory experiences (Vince, 1966; Dimond and Adam, 1972).

2.8.2 Neonatal experiences

The new-born animal faces a profusion of sensory experiences, from which it begins to establish responses to the pleasant and the unpleasant, the vital and the irrelevant. Influenced particularly by its mother and father (if they are still present), its siblings (if any), and others of its species in the vicinity, the young animal begins developing its survival and social skills.

Development is especially shaped by the patterns of stimuli during the neonatal period. Various movements and social relationships are developed and refined as a consequence of stimuli received, often during sensitive periods. The progress of these developments can have a profound influence on an animal's success in adapting to conditions in later life (Wiepkema, 1987). During the period of socialization of puppies (between about 4 and 10 weeks after birth), exposure of pups solely to dogs leads to poor subsequent rapport with humans, while exposure solely to humans can lead to poor relationships with other dogs, to which the pup may show excessive timidity or aggression.

In principle, the raising of chicks without hens, or of puppies in pet shops, or of various types of animals in zoo enclosures would be expected to influence the later responses of those animals. But little is actually known about the effects of early stimulation on animals in these contrived environments where animal welfare problems commonly occur. The influences should not be assumed to be always detrimental. In fact, one might expect animals raised in contrived surroundings to survive better in these conditions in the long term than would animals raised initially in natural environments (Brambell, 1965; Hemsworth, Brand and Willens,

1981; Hemsworth, Barnett and Hansen, 1986, 1987). Studies should be undertaken to provide better information on the development of physiological, behavioural and social regulatory systems. These may reveal how best to equip animals with regulatory systems appropriate to the restricted environments in which they will spend their lives.

2.8.3 Learning and memory

Learning occurs throughout life. Virtually all stimuli, and probably many responses as well, have the potential for being memorized and becoming factors influencing subsequent behaviour and physiological functioning. Developing and memorizing a successful strategy for coping with an environmental disturbance should not only make ensuing exposures easier to deal with but, just as important, lead to the animal having less anxiety at the prospect of encountering such a disturbance.

The consequences of having a memory are not all advantageous. If an animal fails to cope adequately with an imposition, its sensitivity and anxiety may be heightened by recalling this when next that problem looms (Brambell Report, 1965). If the unpleasant imposition was made by a human, handling may become more difficult for the human and more disturbing for the animal which is handled. Just as feedforward control can enable an animal to cope with a stimulus which has in the past been accommodated, so anticipation of a noxious stimulus not successfully countered on a previous occasion may be doubly distressing. Some of the displeasure may be negated if a reward is given simultaneously, and this can be a worthwhile practice when the unpleasant stimulation is beneficial, such as vaccination. Subsequent exposure may then be associated with pleasure rather than displeasure, and hence be accepted more willingly (Beilharz, 1985).

The possibility of animals remembering punishments and rewards has been discussed here as if all species had the memory capacity of humans, but this may not be so (Huntingford and Turner, 1987). The disturbing effect of contemplating a noxious disruption would be expected to be less if the animal had less developed powers of memory, imagination and foresight (Freeman, 1987). But if an animal has a poor memory which makes it fearless of an impending unpleasantness that it has experienced before, it will also suffer the disadvantage of not being able to predict that the unpleasantness will eventually end (Rollin, 1981).

Many studies demonstrate, however, that various animal species can remember events for long periods. One can only be impressed by the demonstration that food preferences of sheep can be influenced by what they, as young lambs, saw their mother eat during brief exposures three years previously (Keogh and Lynch, 1982). In practical terms the findings of Rushen (1986a,b; 1990) are particularly relevant. Rushen assessed the willingness of sheep to proceed down a race after previously experiencing aversive experiences at the end of the same race. As is shown in Fig. 2.6, a sheep would run down the race in a few seconds if it had received no aversive treatment there, but took about two minutes to go down the race if it had been restrained at the far end or if the electrodes of an electro-immobilizer had been attached to it, though not switched on. When the electro-immobilizer was switched on, the sheep took six minutes to proceed down the race.

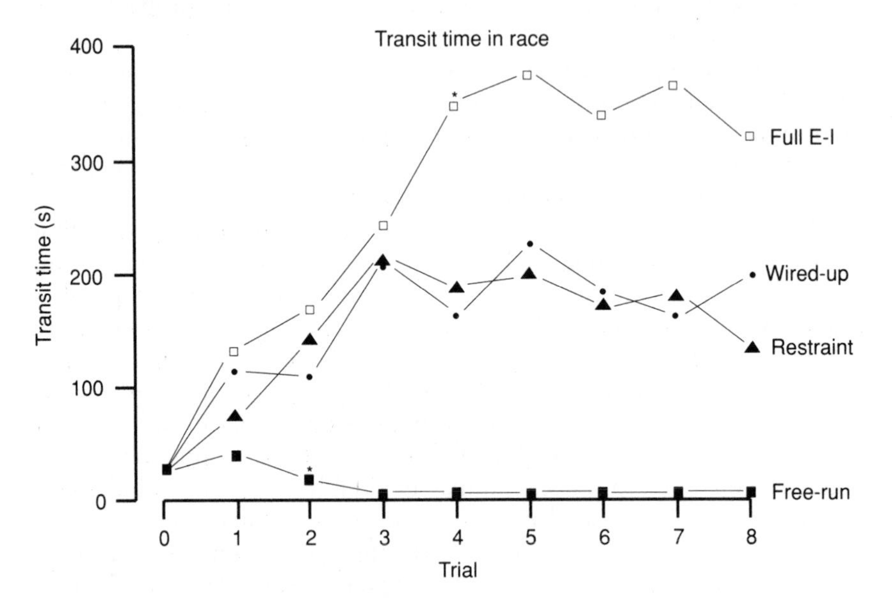

Figure 2.6 Mean time taken by sheep to run through a race in which they were repeatedly electro-immobilized (E-I) or restrained physically (restraint). The sheep in the 'wired-up' group had the electrodes of the immobilizer attached, but the current was not turned on. Asterisks show the first trials at which the treatments differed significantly ($P<0.05$) (after Rushen, 1990).

The reluctance of the sheep to move along the race is a measure of how much aversion the sheep has to its previous treatment. The time interval between the experience and the exhibition of aversive behaviour was a day or two in Rushen's studies, but cows may avoid a crush even months after an uncomfortable treatment there, and dogs can long be reluctant to enter a veterinary surgery where they have had an unpleasant experience.

2.8.4 Lifetime and evolutionary changes

By adulthood, animals have built on their abilities and preferences at birth through a range of subsequent experiences, and have developed various skills. With each new experience, the bank of memories is modified and some slightly different regulatory response may be developed. Thus, during its life, an animal may change its regulatory responses to a given stimulus, because of accumulating experiences, biological and seasonal cycling, and ageing. The change may not always be an improvement (Huntingford and Turner, 1987).

Some of the regulatory responses to the environment which an animal could develop would improve its reproductive capabilities or fitness; others would reduce them. Thus, on the basis of the natural variations associated with age, experience, season and so on, one would expect some genetic selection to take place over many generations for improved adaptation to the environment through refinement of biological and social regulatory systems. If this theoretical prediction were borne out, animals would become progressively better adapted genetically to all the conditions we are exposing them to: sheep to farms, monkeys to zoos, horses to racing, and cats to fast-moving traffic. The slow rate of genetic change, however, makes it unrealistic to expect welfare problems to be solved by such selection.

The probable time-scale for changes of this sort can be guessed. Animals lived for several million years independently of humans, presumably meeting them occasionally when not alert and astute enough to escape their hunting. We have controlled animals through domestication for only a few thousand years, and kept them in close confinement for only a few decades. The influence of the many millennia before domestication will heavily outweigh changes imposed during the last few decades. Some changes have occurred and others will continue to occur, but most characteristics are very resistant to change. Genetic adaptation of hens to battery cages is unlikely to be effected in the near future, even with genetic engineering to accelerate the rate of change. What Rothschild (1986) has said of humans applies equally to other animals: 'We are captives of the genes which enabled us to survive in a prehistoric hostile world.' Evidence that adaptation to the wild is retained despite a veneer of domestication can be seen in the full range of ancestral wild boar behaviour shown by pigs when kept in semi-natural conditions (Wood-Gush, 1988), and the ease with which this species, which seems to require careful management on farms, can revert to feral status and survive in the wild in Australia and New Zealand.

A final but crucial point about the evolution of adaptation to the vicissitudes of the physical and social environment is that a very important part of that evolution has been the development of the complex appreciation of the interactions of an individual with the world in which it lives, which we call feelings. Complex brains, like those of vertebrates, have complex systems for regulating these interactions which are not just the product of automatic responses to stimuli. If an individual has a system of feelings which involves changes in its mental, and perhaps in its hormonal, functioning because a certain kind of body regulation is difficult or because an anticipated event has not occurred, such an individual will have increased fitness in comparison to a genetically different individual which has no such system. The evolutionary advantage of having feelings is considerable (Dawkins, 1990). It would be surprising if animals such as our domestic animals, all of which have an elaborate social organization, did not have feelings similar to many of those of man. A significant consequence of this is that if the various regulatory system components which are manifested as feelings are present in a species, there is a potential for suffering, and that is clearly of great importance both biologically and when considering moral questions.

Chapter 3
Limits to adaptation

The relative constancy of conditions which exists within the body of a healthy animal is achieved by the operation of innumerable homeostatic controls. The control systems react to environmental and endogenous stimuli so as to correct or prevent displacements from the optimal range of conditions. The response is usually proportional to the actual or expected change (Stephens, 1988) and is controlled around a fixed, or sometimes a variable, reference level.

Although biological regulation is occurring constantly, adaptation to stimulation is not always possible. When homeostasis fails 'there is disease and even death', since, as controversially stated by Cannon (1935) and Selye (1976) 'disease is ... a fight to maintain the homeostatic balance of our tissues'. Difficult or inadequate adaptation thus generates animal welfare problems.

Failure to adapt sometimes arises because events occur rapidly, for example, if there is a sudden predator attack or rock fall. More commonly the individual has control systems which, when optimally efficient, either prevent the occurrence of disruptive events or initiate responses effective enough to avoid damage. If the systems do not work well enough, adequate adaptation does not occur, and damage or death ensues. An analysis of what constitutes inadequate adaptation is the subject of this chapter.

Any proportional controller, mechanical or biological, has a limit to its capacity to cope with inputs. An input greater than that which elicits the maximum response causes no further increase in output. Disruption will occur, at least temporarily, in the system as a result of the failure to compensate fully. Adverse stimulation at levels below the maximum can also cause problems of adaptation if the response elicited cannot be sustained indefinitely. Limitations in the capacity of homeostatic control systems can therefore lead to problems of maladaptation and poor welfare.

Stimuli of various sorts bombard an animal continuously. Some are imperceptible, others evoke sensations ranging from trivial to excruciating, with effects varying from inconsequential to life-threatening. What characteristics of stimuli determine whether they will be accommodated by, or exceed the capacity of, a homeostatic control system? The factors of particular importance are the duration and timing of stimulation, and the intensity (Schmidt and Thews, 1983; Becker, 1987) and nature (or modality) of the stimuli. This chapter describes how stimuli – alone, combined, and of various intensities and durations – tax an animal's powers of adaptation. From an understanding of these interrelations and the limits of an animal's powers of adaptation will emerge indications of what constitutes stress, and what is an acceptable level of animal welfare.

3.1 LIMITATIONS OF TIMING AND TEMPORAL ASPECTS OF STIMULUS MODALITY

The interrelations of time, intensity and modality of stimulation are very complex and are depicted in this chapter in simple diagrams to highlight certain features. Somewhat similar diagrams have been presented by Frese and Zapf (1988) as means of illustrating concepts relating to human stress and considering objective measures of such stress.

3.1.1 Changes in frequency

A stimulus such as the sound of a car horn is a relatively innocuous and effective warning signal if heard once a day or once an hour (Fig. 3.1a). At higher frequencies (Fig. 3.1b), habituation may occur and diminish its effectiveness as a warning. We know from our experience that alarms become ineffective if sounded too frequently. On the other hand, if a car horn sounds very frequently, and each sounding is perceived, the stimulation may become irritating, as in the case of persistent car-theft alarms (see discussion of sensitization on p. 40). At such frequencies, there will be an increasing level of response, perhaps reaching a point where the capacity to adapt to the stimuli is exceeded (Fig. 3.1c). For humans, the experience of car horns constantly sounding in dense traffic can be disorienting. A dog crossing a busy road may also become disoriented because multiple inputs result in a sensory overload that its control systems are not able to accommodate.

Another factor determining the impact of pleasant or unpleasant stimuli is the rapidity of adaptation of the response to a stimulus (Fig. 3.1d). This can vary widely depending on the nature of the stimulus and its biological significance. Consider two examples of visual stimulation. The physiological reaction to the sight of a detailed geometric pattern may pass within milliseconds or seconds. On the other hand, the effects of seeing a predator may last for hours. Though the complexity of the visual stimuli may differ little, the consequences will contrast dramatically. Repeating exposure to the geometric pattern would cause an inconsequential reaction whereas sighting the predator again may cause alarm. However, responses to even horrifying stimuli may habituate in time if they are not directly threatening; witness the passivity that people can develop to

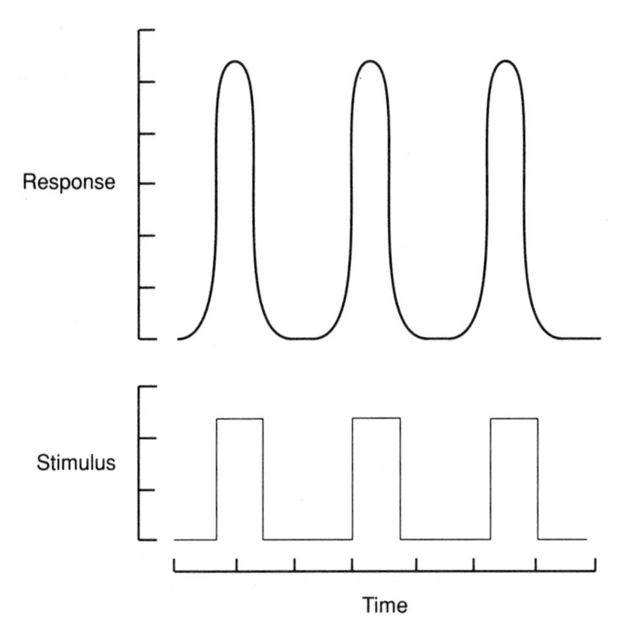

Figure 3.1(a) Successive responses to stimuli without habituation.

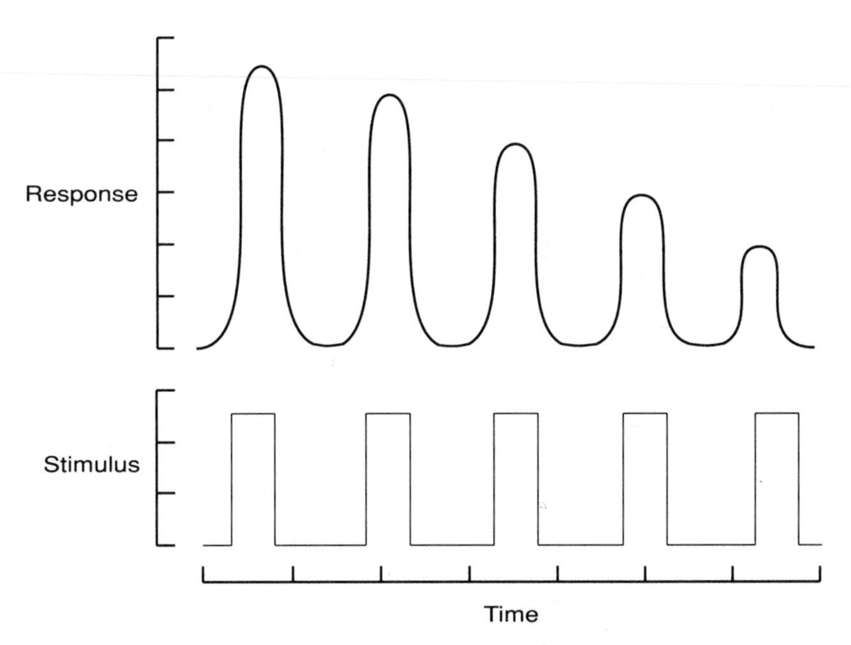

Figure 3.1(b) Successive responses to stimuli with habituation.

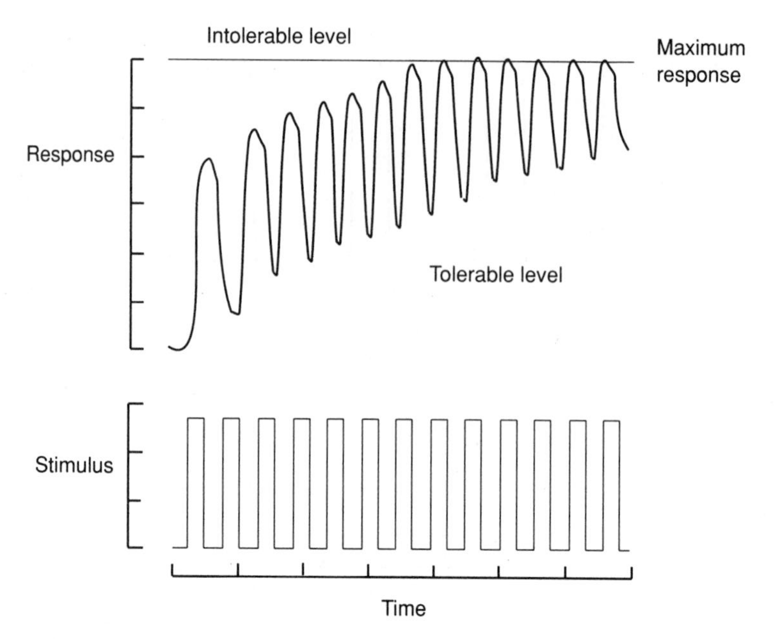

Figure 3.1(c) Successive responses to stimuli with sensitization.

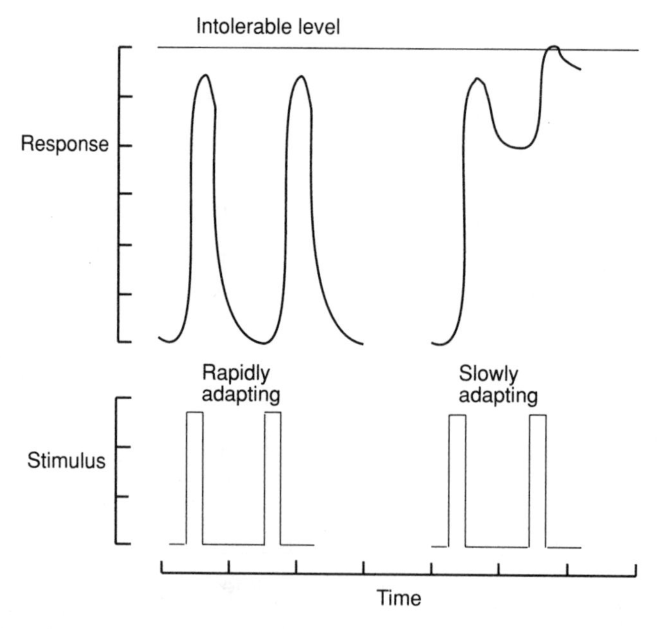

Figure 3.1(d) Responses which adapt before the next stimulus occurrence or which have not adapted by the next stimulus occurrence.

wartime horror. A difference in response thus arises not solely from the nature of the stimulus but also, in part, from the interpretation of the signals in the light of past experience.

Habituation is not a result of simple adaptation of receptors or of fatigue in effectors such as muscles (Brown, 1991). Nor is it the result solely of fatigue in other parts of the nervous pathway between receptor and effector; habituation must involve complex processing of multiple inputs. Sokolov (1960) described the habituation of the startle response of dogs on repeated presentation of a tone, and the reappearance of the response when the tone was varied in pitch or quality. Broom (1968) found that young domestic chicks exhibited a startle response when a small light in their pen was switched on for 10 s, then left off for 20 s.

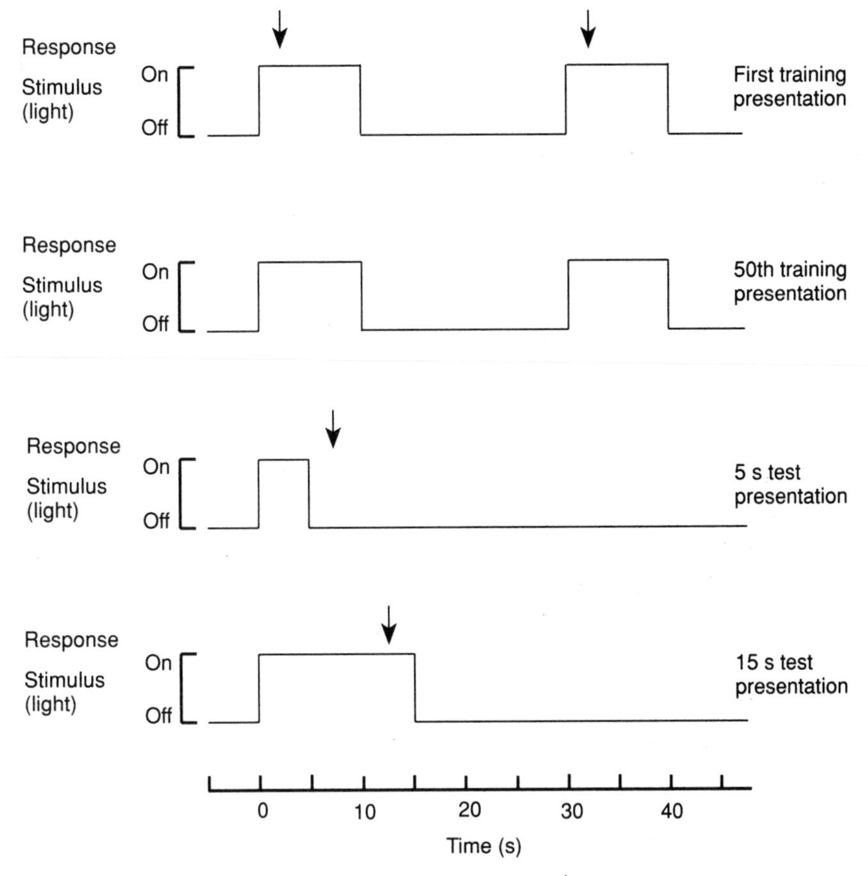

Figure 3.2 Young domestic chicks responded (↓) when a small light was switched on for 10 s in their home pen. This response habituated well before the 50th presentation but returned if the light went off too early or stayed on too long (data from Broom, 1968).

With repeated presentation the response habituated but it reappeared if the duration of illumination was changed. Furthermore if the light went off after 5 s there was a reaction at that time. If the light stayed on beyond 10 s there was a response just after 10 s when the light should have gone off (Fig. 3.2). These results show that the habituation in both sets of experiments was for specific timing of illumination and that the animals must have had precise expectations about the stimulus input, which they matched with the actual input. The waning of a response to a stimulus can obviously involve complex cognitive processes.

A quite separate reason for different rates of adaptation can be dissimilarities in the nature of the stimuli. Most stimuli have very brief effects on the sensors. However, some that are transiently noxious, such as cuts, scratches and bruises, may have long-term residual effects related to the extent of tissue damage. This could be interpreted either as slow adaptation or as a stimulus of long duration. The effects of prolonged stimulation will be considered separately below.

3.1.2 Changes in duration

A sustained painful stimulus is clearly a greater imposition than a brief painful stimulus. However, the effect is not simply related to stimulus duration.

The effects of a brief stimulus (Fig. 3.3a) usually fade away at a rate depending on the characteristics of the stimulus and the animal, as described in the preceding section. For some sustained stimuli, a similar adaptation may also occur (Type 1), for example, after a saddle is put on a horse. The influence of the stimulus abates by adaptation despite its continued application.

Other long-lasting stimuli can have an increasing impact with time (Fig. 3.3a, Type 2). The stimulus in this case may be tolerated if it lasts only briefly but, with the passage of time, the response by the animal increases. The animal becomes more and more sensitive despite the constancy of the stimulus. Such an increase in response to continuing or repeated stimulation is called **sensitization** (Section 2.2.3). By such progression, a stimulus that is initially tolerable may eventually reach an intolerable level. The biting insects, quoted in Chapter 2, or the sound of a pneumatic drill are repetitive stimuli to which people may become increasingly sensitive. What begins as a mild irritant can eventually become intolerable. Another example of this is the 'dripping water torture', where a stimulus that is quite innocuous when applied in isolation (namely, a drop of water on the forehead) is said to become sufficiently noxious to send a person crazy when applied repeatedly over hours and days. Interestingly, Lazarus and Folkman (1984) suggest that humans are less tolerant of sustained petty disruptions than of major tragic events in their life.

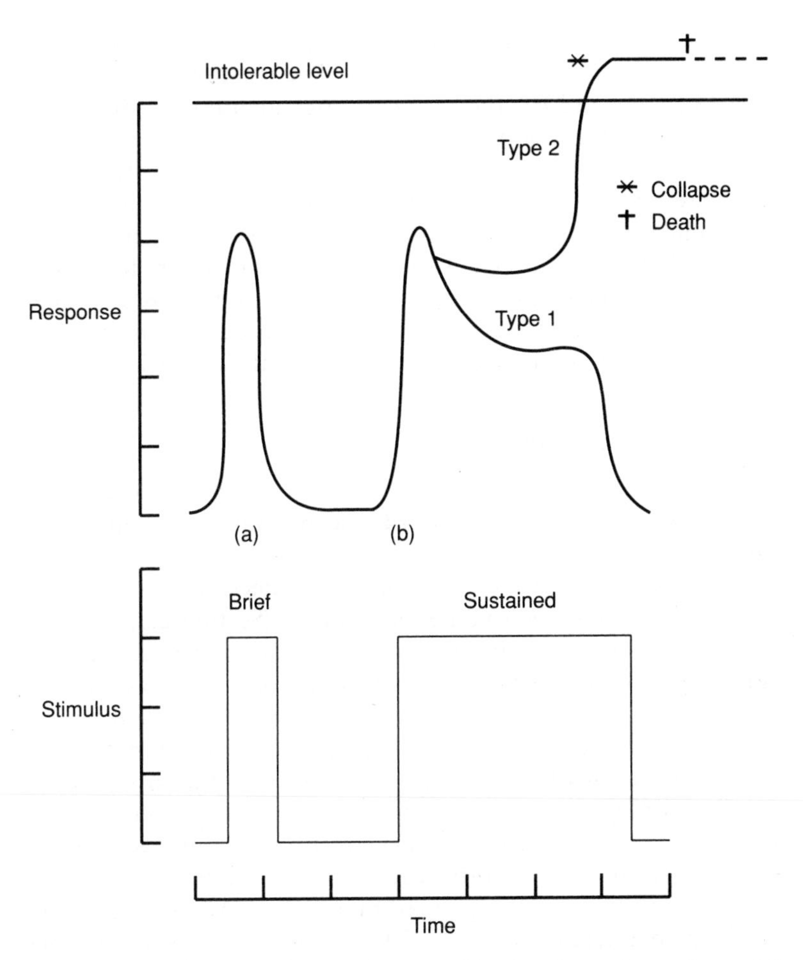

Figure 3.3 Schematic relations between (a) a brief stimulus and a rapidly adapting response, (b) a sustained stimulus and a response which adapts (Type 1) or is sensitized (Type 2).

If a stimulus exceeds the threshold of tolerance (Fig. 3.3b, Type 2) and an animal's homeostasis is progressively disrupted, the effect on the animal depends directly on the duration of stimulus. Only brief exposures can be endured; prolonged exposures will lead to collapse and death.

3.1.3 The impact of novelty

Responses depend on previous experience of the stimulus. Novel stimuli cause an animal to become alert. If the stimulus is of little significance, the animal may show some curiosity but this will wane if the stimulus is repeated. If the stimulus is of great significance, the response will be one of continuing strong interest, or even alarm. The reaction to subsequent stimuli will then depend on the characteristics mentioned previously: intensity, duration and modality.

Novelty enhances the impact of a stimulus on an animal. A noise may generate little reaction from an animal that is familiar with it, but provoke alarm in an animal which has not heard it previously; that is, the second animal's threshold for response will be lower. Lack of exposure to a varied environment will result in the animal showing alarm responses to quite mild stimuli. As a result, stimulation that seems to be within everyday experience may push inexperienced animals beyond their capacity to cope.

3.1.4 The value of forewarning

Forewarning that a stimulus is coming can help an animal prepare for unpleasant consequences. Feedforward mechanisms (Section 2.6.2) prepare animals for either environmental or social change.

A situation in which preparation for unpleasant effects may be necessary occurs when regulatory systems come into conflict (Section 2.5.4) and an animal has to remain in conditions that impose on it in some respect. When animals have to tolerate adverse conditions, such as low air temperature or high predation risk to obtain food, they benefit greatly from previous knowledge and presumably prepare for the adversity, since deciding to forage necessarily involves prior assessment of the environment.

In man, forewarning can result in acceptance of a painful treatment with the expectation of relief from pain or other suffering. A human may accept an injection of an anaesthetic drug before dental treatment, or of an antibiotic drug as the cost of being cured of a disease. A domestic animal may be helped to tolerate such 'costs' by its owner's reassuring voice encouraging it to accept stimuli against which it would normally react.

3.2 LIMITATIONS OF INTENSITY

3.2.1 Changes in intensity

Varying the intensity of a stimulus may appear to influence an animal because this causes a greater or smaller number of impulses to be transmitted along nerves to the central nervous system; but there is more to it than that.

To illustrate, consider the difference between the visual stimulation resulting from a dull and a bright beam of light. Contrast this with the difference between stimulation by a bright, narrow beam and by a floodlight. In both situations there are differences in the impact of the stimuli, but the difference in the first case is in the intensity of the applied stimulus and, in the second, in the area over which the stimulus is applied. The light from the dull and bright beams may impinge on exactly the same cells of the retina, though the bright light activates those particular cells more than the dull one. The floodlight may affect individual retinal cells to the same extent as a bright beam, but its greater effect comes from the larger number of cells affected and hence the increased number of activated channels to the brain. Other differences of importance in animal welfare are those related to spatial and temporal summations of temperature, pressure, pain and other stimuli.

In some situations, the same stimulating effect can be brought about by stimuli in various combinations of intensity and area, as shown schematically in Fig. 3.4. The greater is the total stimulation of the animal, the greater will be the animal response.

In practical terms, the contributions of density and area to total stimulation are especially important in assessing the effect of local, compared with generalized, trauma in an animal. Parts of an animal can be grossly disturbed locally, for example by infection or trauma, yet constitute only a moderate handicap to the animal. At the affected site, measures of damage, such as the frequency of pain-sensor discharge or the amount of degraded tissue, may indicate a serious state. Yet if the site affected is small, the overall effect on the animal may be inconsequential. Conversely, a stimulus of much milder intensity acting over a large area may constitute a handicap that leads to debilitation and death. The practical implications of this are important in cases of chronic infestation with parasites, or the simultaneous imposition of numbers of minor stimuli any one of which might be imperceptible, but which collectively cause drastic disturbances to animal function.

3.2.2 Hazard avoidance and lethal limits

A stimulus may be so intense as to constitute an immediate threat to life. When an animal encounters such a threat, its response may be dominated by one of the physiological mechanisms that have apparently evolved to cope with such emergencies. These include the withdrawal reflex, the orientation reaction, the alarm response, and the tendency to flee or 'freeze', i.e. stay still. High intensity heat, light, noise, or pain activate powerful neurally based reflexes to withdraw the affected part of the body from the noxious stimulus. In emergency situations all of these responses minimize damage and promote survival, but the inflexibility of the response can constitute a handicap. For example, reflex flight from a competitor without planning may exacerbate the problem. Freezing can be an appropriate and efficient response to threat but it has also been reported to disorganize other escape behaviour (Stephens, 1988): animals stopping while crossing a busy road are sad examples.

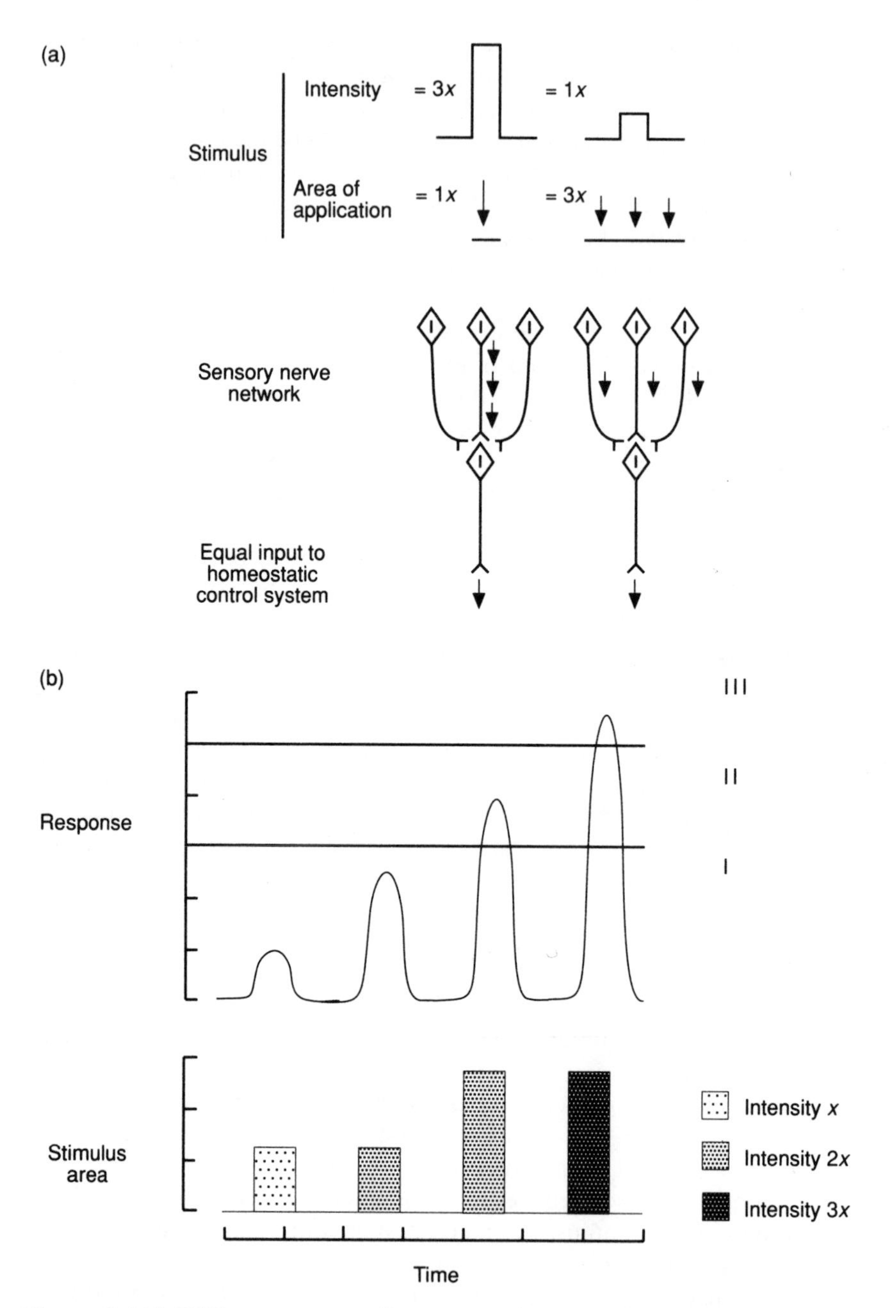

Figure 3.4(a) Different patterns of nerve stimulation resulting in equal inputs to homeostatic control mechanisms. (b) Different combinations of stimulus intensity and area which result in responses which are associated with: (I) easy coping, (II) coping with difficulty, or (III) failure to cope.

In extreme cases, stimuli can cause irreversible damage very quickly to an animal. Burns and traumatic wounds may be so severe that once inflicted, an early death is the only possible outcome. Alarm and emergency responses may help animals avoid or minimize the effects of some hazards. For instance, toxic substances in the diet are hazards which many individuals encounter during their lifetime. The factors which minimize the likelihood of them having harmful effects include detoxification, especially by the liver, and behaviour patterns such as avoidance of novel foods, taking only small quantities of such foods, and learning to associate adverse effects with a particular food. The adaptation of an animal to an environment containing poisons will be limited by its capacity for detoxification, and by any failure to limit input or learn quickly that an ingested substance is harmful (Broom, 1981b).

3.3 THE SIGNIFICANCE OF DIFFERENT MODES OF STIMULATION

The colonization by animals of virtually every part of the world – deserts, swamps, city centres and Arctic tundra – indicates their wide tolerance of environmental conditions. There are, however, obvious species differences in sensitivity to various factors such as dehydration, noise, temperature, and so on. Some animals are acutely sensitive to cold, others largely unaffected by it; some need food frequently, others easily cope with periods of starvation. Adverse circumstances are not the same for all species, and this must be borne in mind when considering how to minimize stress and improve welfare. Ultimately a unique assessment is required for each situation, as described in Chapter 9. Some generalizations can, however, be made.

Two potentially noxious stimuli imposed together will generally evoke a greater response than either alone. However, additivity (a simple summation of the responses) cannot be assumed, for the combined response may be greater than the sum of the components. An occasional rustle in the undergrowth or a distant wolf howl might not result in any response from a deer, whereas both together might precipitate flight. On the other hand, the combined effect of two factors might be less. Frisch (1981) has shown that cattle living in tropical climates eat less and have a resultant fall in metabolic heat production which allows them to tolerate the hot climate better. Hence providing less food for animals in environmental temperatures above normal can lead to less of a problem for the animal than would imposition of either food deprivation or heat load alone. In fact animals adapted to such conditions over many generations have an inherently lower voluntary feed intake, together with an improved heat tolerance.

3.4 INTEGRATING TIME, INTENSITY AND MODE OF STIMULATION

The stimuli impinging on an animal vary in time (i.e. frequency and duration), intensity (i.e. density and area), mode (i.e. visual, gustatory, emotional, etc.), and degree of novelty. Before discussing how the process of integration might be achieved, we can outline the relatively simpler processes that presumably operate to integrate stimuli with different temporal and intensity characteristics. Figure 3.5 illustrates schematically how a sequence of stimuli of similar modality but varying duration and intensity could combine to constitute a lethal imposition, even though each one by itself is only moderate in its effects.

Figure 3.5 could be used to illustrate changes in a single animal in a group of pigs, in which a dominance hierarchy has been established, in response to a particular sequence of events. At points 1 and 2, gestures of challenge are made by other animals in the group, and the visual signals evoke transient responses which decline rapidly. Stimulus 3 arises because the animal is removed, somewhat roughly, from the group and put alone in a crate; the novelty of the surroundings elicits an increased response. Although the animal begins to habituate to this new situation, before the response has appreciably subsided, the animal is driven by an unfamiliar and unsympathetic stockman to a set of weighing scales, at point 4.

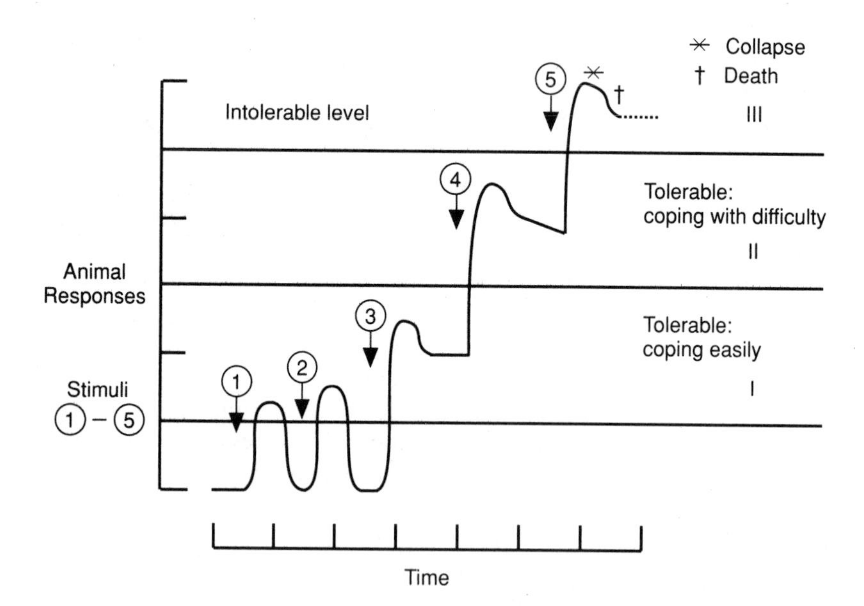

Figure 3.5 Responses to a series of stimuli which, individually, have moderate effects but which can be lethal in combination.

Domestic pigs, especially those of some breeds, are severely affected by handling and may suffer cardiac problems as a consequence. When the same handler returns and begins loading the animal onto a truck, at point 5, the experience adds to the animal's previous unpleasant experience and induces a cumulative effect such that the animal is at risk of collapse.

In this example, we can easily appreciate the nature of the stimuli because they are visual or physical experiences, or recollections of such factors. In reality, stimulations are likely to be of various modalities. An animal may see a predator, hear a call from its offspring, experience a parasite biting its skin, have a sensation from its empty stomach, or be frustrated as a consequence of some restrictive aspect of its environment. All of these, and other stimuli as well, may converge on the animal's sensorium simultaneously.

Are the assorted stimuli additive? The combined effect of a number of stimuli is likely to be more than that of any single stimulus. It is also possible that a combined stimulation could overwhelm an animal, even if each component stimulus could easily be accommodated. But, for the reasons enunciated above, additivity cannot be assumed.

Is it possible then to predict the integrated effect? The practical answer is again, no, for several reasons. Stimuli vary in their sensory impact on an animal, in the responses they evoke, and in the extent to which they are remembered, and thus influence subsequent exposures. Some stimuli elicit a reflex nervous response, others a hormonal response of slow onset. Some species have acute sensitivity to particular stimuli and relative insensitivity to others. Beyond appreciating that environmental and endogenous stimuli are collectively and cumulatively affecting each animal, the task of attempting to predict the integrated effect is too complex.

Before proceeding with discussion of integrated stimuli, it is worth noting the proposal of Selye (1950) that the combined impact of noxious stimulation is indicated by the output of glucocorticoid hormones from the adrenal cortex. Unfortunately Selye's concept does not accurately reflect the problems of integration, for it has been shown to be sometimes incorrect, as will be discussed more fully in Section 4.1.1.

3.5 THE CONCEPTS OF TOLERANCE AND COPING

Figures 3.4b and 3.5 depict the range of stimulation and response that an animal can tolerate, and this is divided into three. Responses in the first range (i.e. those below the first level marked on the diagrams) are to a stimulus which elicits a measurable response but which is tolerable indefinitely. For instance, people cope with such impositions as moderate heat or cold exposure, partial loss of sight or loss of some teeth; humans or other animals can bear easily minor scratches or bruises, a low level of parasitism, or low-level social abrasion. These impositions may have a measurable effect on the individual concerned but, if the suggested grading is valid, they will have an effect that is readily tolerable and has no detectable detrimental effect in the long term. In fact, it will be argued later, that some stimulation early in life may be essential for effective later adaptation.

Appreciably greater impositions on an animal may also be tolerable indefinitely, but induce temporary penalties of displaced homeostatic control and disturbed behaviour. These responses are in the second range and would often arise from a stimulus that causes pain but not permanent damage. This is the most widespread category of situations that adversely affect welfare. In human terms, such a situation might occur when a man has difficulties at work over several months, during which he develops stomach ulcers, finds social interactions troublesome, develops neurotic behaviour and succumbs more readily than usual to viral infections. He survives, and there may be no long-term detrimental effects, but coping has clearly been difficult for him during this time. In fact, most people would consider that these are circumstances in which the welfare of the individual is considerably worse than normal, and would try hard to avoid them.

The most serious impositions on an animal's homeostasis are stimuli that elicit a response in the top range, which can be tolerated only for strictly limited periods, and otherwise cause death. Within this range, stimuli are tolerated for a finite time – perhaps a minute, perhaps a month – before the animal succumbs. Figure 3.6 depicts the range of responses to stimuli in relation to coping and tolerance. The upper limit of the top range is the maximum response possible by the animal. Should the stimulus strength exceed that which elicits the maximum response, some of the disturbance will, by definition, be uncompensated. Incompletely compensated disturbance of a homeostatic system will cause progressive displacement of physiological variables outside the range necessary for survival. The animal will succumb unless corrective measures are taken.

By proposing that there are varying levels of tolerance we are also introducing the notion that there is an inverse relation between strength of stimulus and tolerance time. That is to say, stimulations that are severe will be tolerated only briefly. A painful but localized injury such as a deep knife cut, or a traumatic event such as the death of an acquaintance are disturbing events that would be even more distressing if their effects were not relatively transient. More moderate stimulation, such as from a viral infection or social antagonism, is tolerable for longer periods, but if it continues indefinitely, the person or animal will be handicapped by the excessive stimulation and may die prematurely. The mildest effects of endogenous or exogenous disturbance can be coped with without a biological penalty, and may perhaps even confer a biological advantage in terms of lifestyle, production or reproduction; for example, when experience of minor social altercations prevents the subsequent occurrence of more severe social problems.

An important concept is introduced in the relations depicted in Fig. 3.6 – the concept of **coping**. In the scientific literature the ability to tolerate different degrees of stimulation, particularly noxious stimulation, is embodied in the concept of coping. Coping is defined as having control of mental and bodily stability (Fraser and Broom, 1990). In an account of human adaptation to various stimuli, Lazarus and Folkman (1984) suggest that it is the extent of the ability to cope that ultimately determines whether the individual survives in unfavourable conditions.

Referring again to Fig. 3.6, we could propose that animals experiencing level I stimulation are able to cope indefinitely and there is no welfare problem. Those exposed to level II stimulation may be coping as individuals, in that their fitness is not reduced, but they do so only with difficulty. Level III stimulation results in reduced fitness and failure to cope.

A further brief explanation should be made of the interrelation between tolerance and coping. When an animal is under a moderate degree of continuous stimulation, but nonetheless coping, as in Fig. 3.5 after the third stimulus, its capacity to cope with subsequent stimuli could be

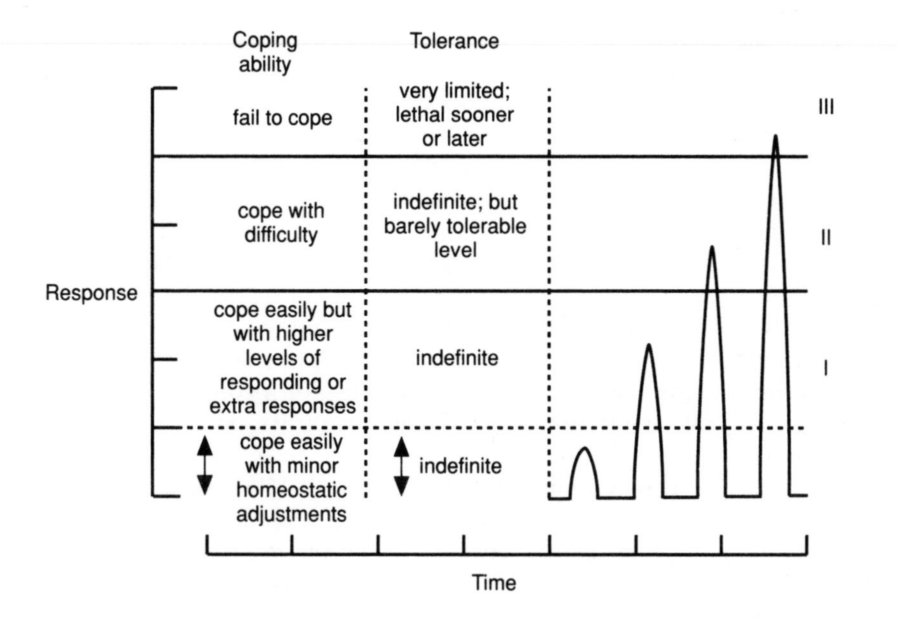

Figure 3.6 Levels of responses to stimuli in relation to coping ability and the extent to which these stimuli and responses can be tolerated.

reduced. In Fig. 3.5, the capacity for coping with future difficulties after the third stimulus is only about half that existing in the unstimulated animal. This is only one possible scenario as it could be that by switching to an alternative coping method, e.g. from active flight to freezing, the individual could enhance its efficiency of coping.

3.6 VARIATIONS IN PATTERNS OF ADAPTATION

3.6.1 Differing rates of adaptation

Disturbances of an animal's homeostatic regulation induced, for example, by changing its feeding site, thermal regime or social grouping, take varying periods to be accommodated. If the feeding site on a farm or in a zoo were changed, the visual stimulus would be instantly recognized by the animal. The animal's adjustments to the change would be rapid as long as there were no other social readjustments to be made, say, by being forced into the proximity of a rival. If, on the other hand, the feeding place of an animal was altered because it was being transported by truck, the modified visual stimulus would soon be accommodated, but the animal would need to develop new skills of balance and of coping with noise, vibration, and other physiological and chemical problems. Adaptation of physiological systems to this more complex environment would be slower.

Individual animals differ in their acceptance of change for several reasons. First, each animal has an established place in its various social systems, such as its territory or hierarchy. Associated with that position are patterns of where and when to feed, with whom to associate, and so on. Disruption of established patterns may disturb some individuals of a group more than others. In humans, individual differences in response to disturbance are well recognized (Lazarus and Folkman, 1984) and have been ascribed to differences in 'coping'. Such individual differences probably exist in other species, though they are not so easy to discern, at least partly owing to our difficulties in interpreting their behaviour. The problems involved in breeding from dominant hens, which are disinclined to adopt the submissive sexual crouch necessary for mating (Kilgour and Dalton, 1984), illustrates how the challenges of adaptation are unique for each animal.

A second reason why individuals differ in adaptability to disruption is the association between sensitivity to disturbance and biological rhythms, particularly the ovarian cycle. Females may accept environmental change more readily at one stage of their ovarian cycle than another. Sexual activity can also be strongly seasonal and hierarchical in males, such as deer. The effect of disturbance on stags will therefore depend greatly on the time of year and the social position of the animal.

Each animal has a unique set of lifetime experiences upon which the process of accommodating change depends. Thus, there are striking differences among individuals in their level of behavioural (Syme and Elphick, 1982) and physiological responses (Fell and Shutt, 1986) to environmental manipulation.

Finally, there are species differences in the readiness with which animals adapt to environmental disruptions. The ease with which a species becomes domesticated is a special case of this (Hemmer, 1990; Budiansky, 1992). Certain species, for example bears, are slower to adapt to life in zoos than are others. They may be unable to adapt, and show abnormalities of behaviour, physiology, reproductive ability or disease susceptibility. In contrast, some ungulates appear to adapt quickly and quite adequately to good zoo conditions. The difference is presumably a result of adaptation to the normal ecosystem in which the species lives, and it exists even when equal attempts have been made to promote domestication. This is difficult to show conclusively with those species which have experienced human intervention for many thousands of generations, for they will have altered genotypes as a result.

To represent diagrammatically the variability in tolerance of disturbing or noxious stimulation, each individual would really require its own set of critical values of response. Furthermore, the scales on the diagrams would need to be altered with ovarian cycle, season, social status, and so on. In practice, obviously only the average expectation for each species could, at best, be forecast.

Working with the mean response of groups, rather than with that of individuals, introduces new problems. If it is true that some animals in a group suffer appreciably more than average, and thus are more likely to fall below objectively set response standards, then standards should theoretically be set for every individual, rather than for groups. Clearly this is impractical except for very small numbers, such as animals in zoos. The problem raises the question of what proportion of the animals in a group should have good welfare. This question is of both biological and ethical significance.

3.6.2 Hypersensitivity

The levels at which stimulation becomes intolerable can alter with physiological and seasonal rhythms, and also depends on personal experience. They are set uniquely for each individual. Hypersensitivity is a further factor affecting the responses of individuals to environmental change. A possible physiological basis of hypersensitive responses has been discussed in Section 3.1.2.

In terms of the operation of an input–output model of physiological regulation (Fig. 2.3b), the cause of hypersensitivity could be considered as a lowering of the set-point of the control system, or as an increase in the gain of the system. The result is that a given stimulus elicits a greater response than expected in such an animal. In circumstances of increasing stimulation, intolerable levels of response are reached more rapidly in a hypersensitive animal, and if these continue the animal will succumb and die sooner.

3.6.3 Hyposensitivity and stress-induced analgesia

Reduced sensitivity to stimulation, or hyposensitivity, has been the subject of special study because of its relation to pain relief, or analgesia. Testing whether some animals tolerate noxious stimulation better than others, and whether it is because they have lower sensitivity, is fraught with methodological difficulties. Nonetheless, considerable evidence is available that exposure of animals to noxious stimulation can be associated with some suppression of the pain response because substances that have an analgesic effect are released within the central nervous system (CNS) and maybe elsewhere in the body. These substances, the enkephalins, endorphins and dynorphin, are chemically related to opioid drugs used as analgesics (Amit and Gallina, 1986; Becker, 1987). They occur in various CNS pathways, and their action in the brain can be blocked by various substances including naloxone.

Agents other than endogenous opioids may also be involved (Devor, 1984; Lewis *et al.*, 1984; Amit and Gallina, 1986). Opioids commonly act over long time-spans, but some analgesia is produced quickly. Furthermore, stimulation of sites in some parts of the brain, such as the *locus coeruleus*, inhibit pain rapidly. The non-opioid systems for stress-induced analgesia are short term, and may operate through other mediators such as vasopressin which is known to be released following injury (Anderson *et al.*, 1989). The combined short- and long-term systems appear to act as specific adaptive mechanisms to generalized excesses of stimulation. The best known examples of such analgesia are the reports, by those in battle, of limb loss or other very severe injury, without any sensation that this has happened. In other experimental studies, Komisaruk and Larsson (1971) found that there was a substantial suppression of responses to noxious stimuli in female rats when they were subjected to mechanical pressure on the uterine cervix, and many authors have demonstrated that the delay before rats withdraw their tail from a gradually increasing pinch or heat source is much increased by a variety of unpleasant experiences beforehand, such as inescapable foot shock or cold water swims (Akil *et al.*, 1976; Bodnar, 1984).

Free-living animals repeatedly face situations in which noxious stimulation is unavoidable, for example, when defending territory or offspring, or foraging in hostile climates. When caught and injured by a predator, it may be best for an animal's chances of survival not to show an obvious pain response. Animals sometimes appear to accept a variety of unpleasant effects by adopting responses which allow them to carry out activities essential for health or survival (Houpt and Wolski, 1982). Resignation to a situation, perhaps with self-narcotization using endogenous opioids, may allow an individual to survive and reproduce. However this must be considered an extreme form of trying to cope.

3.7 OTHER FACTORS AFFECTING ADAPTATION

The factors detailed here, like some of those mentioned earlier, are examples of the consequences of situations in which an animal lacks control of its interactions with its environment. Adaptation is clearly more difficult, or indeed impossible, if the control mechanisms cannot operate properly.

3.7.1 Lack of stimulation

In natural conditions, animals are constantly stimulated by changes in their physical and social environments. Where animals are brought under closer environmental control, on farms, in zoos, or in people's homes as pets, the levels of some of the components of stimulation are reduced, while others are increased. The reduced environmental stimulation is usually planned to minimize the adverse effects of variability in temperature, food supply, and so on. Paradoxically, although some adverse effects may be reduced, others can increase considerably. The net effect of attempting to reduce stimulation can actually be greater disturbance of the animal.

The most cogent explanation of this response is that animals have expectations of the consequences of different types of activity such as foraging, social interaction, and so on. Where these do not materialize, perhaps because the environment is controlled by humans, the animals themselves are not able to utilize fully their own array of controlling procedures. Some animals respond to lack of stimulation with apathy, a response apparently associated with the lowest level of stimulation. Other animals, instead of becoming apathetic, may contrive to replace programmed requirements with stimuli of their own making, so that these appear as repetitious activities, or stereotypies. Not being responses to specific stimuli, such stereotypies can be manifest not only as purposeless routines, but as behaviour which damages the animal itself or others in its vicinity. There are certain parallels in humans who experience lack of stimulation, or underload, at work. Such people can develop psychological disturbances due to failure to adapt (Cooper and Payne, 1988).

The diagram developed earlier to illustrate the relation between level of stimulation and level of response (Fig. 3.4) could therefore be modified to show that with sustained minimal levels of stimulation, that is during sensory deprivation, the response is not minimal. It is often appreciably elevated (Fig. 3.7), with adverse effects.

A second, equally serious, consequence of lack of environmental stimulation is a loss of the capacity to adapt to new environments. The mechanisms of adaptation appear to require repeated use and reinforcement, otherwise the ability to adapt is reduced. Domestic animals usually cope with a range of minor disruptions in their management routine (Brambell, 1965), though cope less well with rarer experiences such as transport which may not previously have been encountered (Kilgour and Dalton, 1984). Wild animals may not cope well with even the mild regular management procedures associated with domestication (Brambell, 1965).

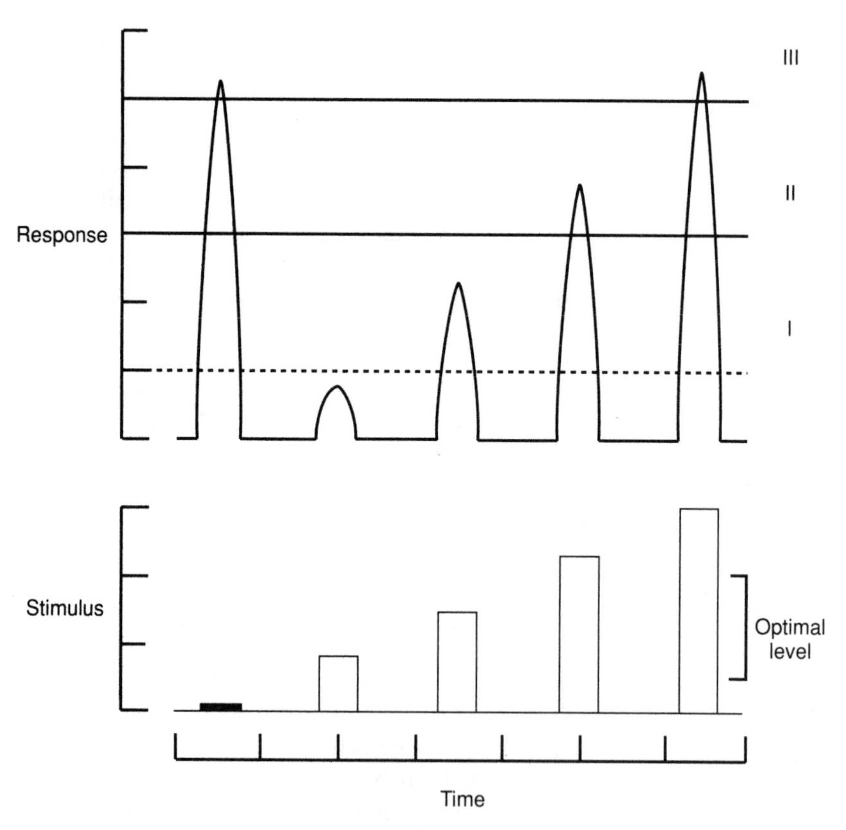

Figure 3.7 Responses to different levels of stimulation. If the level of stimulation is too low, responses can be high. Sensory deprivation or barren environments can lead to poor welfare.

3.7.2 Unpredictable stimulation

Response to stimulation is always likely to be more effective if there is some accurate anticipation of what is likely to happen. Mechanisms to achieve this exist as feedforward controls (Section 2.6.2). When there has been no prior experience of a situation, there can be no anticipation of what might be encountered. The novelty is then additionally disturbing and commonly prompts an increased response (Broom, 1985).

A variation of this problem arises when a pattern of stimulus and response has been established, for example in relation to feeding times or social groupings, but is then unpredictably disrupted. Animals encountering such unpredictable stimulation can become aggressive. Carlstead (1986) presented food to pigs kept in groups and, at the same time, a bell was sounded. If the bell and food presentation were suddenly changed so that they were no longer associated in time, the pigs showed much more aggression than they did during the period when the stimuli were associated, and more than did pigs which had been presented with the bell and food randomly throughout the study. When such situations continue, animals begin to show signs of maladaptation. For example, rats that do not receive food when expected have heightened adrenal responses (Levine, Goldman and Coover, 1972).

In terms of body regulation, animals show a greater response to a given stimulus if they cannot predict it, so it is treated largely as if it were novel (Section 3.1.3). Unpredictable stimulation, or the absence of expected inputs, means that the animal has less control and is not able to regulate effectively. The responses to unpredictable, rather than predictable, stimulation are heightened, and animal function is more disturbed (Broom, 1985).

3.7.3 Frustration of behavioural output

Adaptation can also be disrupted if most of the factors required to effect a response are present, but critical ones are missing. In some circumstances, a genetically established reflex or learned response, such as to feed, mate or escape, is appropriate, but one, or a few, of the essential factors are absent and the animal is thwarted in its efforts to carry out the activity. If the levels of most of the causal factors which promote a behaviour are high enough for the occurrence of the behaviour to be very likely but, because of the absence of a key stimulus or the presence of some physical or social barrier, the behaviour cannot occur, the animal is said to be **frustrated** (Broom, 1985).

If, under these circumstances, a response cannot be completed, the animal may direct its energies into another activity, not uncommonly into aggression against nearby animals. Duncan and Wood-Gush (1971, 1972) accustomed pairs of hens to being fed from a dish in a cage. If the dish was covered with perspex, so that the hens were frustrated in their attempts to get at the food, the hens showed a substantial increase in stereotyped pacing (Table 3.1) and an increase in aggressive pecking by one of the pair (Table 3.2). The response is similar in some of its physiological and behavioural characteristics to that of animals attempting to adapt to noxious stimulation. Frustration can only be measured by the level of some non-specific responses. Like unpredictability of stimulation, frustration disrupts established patterns of adaptation to environmental change.

Table 3.1 Effects of frustrating hens on stereotyped pacing (from Duncan and Wood-Gush, 1972)

	Mean number of stereotyped pacing routines in 30 minutes
Hen deprived of food then fed	13.3
Hen not deprived	18.7
Hen deprived of food then food under perspex cover	161.0

Table 3.2 Effects of frustrating hen by food deprivation on attacks on subordinate hen (from Duncan and Wood-Gush, 1971)

Hours of food deprivation	*Median number of attacks on subordinate hen in 20 minutes*
2.5	3
5	11
7.5	18

When animals attempt to respond to stimuli which are associated with very aversive events, but are repeatedly unable to prevent the event, they may stop responding entirely to events in the world around them. This condition occurs in rats following repeated inescapable shock, and was called 'learned helplessness' by Maier and Seligman (1976). It was suggested that the rats learned that there was no way in which they could prevent themselves from being shocked, and so underwent various emotional and motivational changes which made it difficult for them to learn any new associations. It is possible that experiencing the uncontrollable shocks resulted in some opioid action in the brain which contributed to poor performance in other learning tasks (Maier and Jackson, 1979).

Chapter 4

Stress and strain, welfare and suffering

Up to this point, disturbances to the systems regulating an animal's life have been discussed in terms of stimuli initiated by sensory receptors both inside and outside the body, or arising from internal body changes which affect the brain by routes other than via sensory receptors, or arising from brain function without other input. These are interpreted in relation to previous experience and, hence, result in causal factors which are inputs to the decision-making centres producing regulatory responses and contributing to homeostasis. This mechanistic analysis can now be extended to the topics of central concern in this text: stress, animal welfare and suffering.

In most situations, the regulation of body function during disturbances is achieved by simple actions which cause no appreciable disruption to the animal and which may be achieved in a largely automatic way. However, normal regulatory mechanisms cannot deal with some impositions of the environment on individuals. Some necessitate extra physiological or psychological resource commitment, including extra energy consumption for adaptation. These responses may be brief or long lasting, but can often be tolerated indefinitely without adverse effects on an individual. Of greater impact are environmental impositions which are sufficiently noxious, intense or prolonged to make even the maximum response inadequate to maintain homeostatic balance, and so the individual experiences adverse effects. The most extreme are those hazardous stimuli that occur so quickly as to virtually preclude a physiological or behavioural response, that is, the ones that are rapidly lethal. In this chapter, these different levels of stimulation, response and consequence will be used to define degrees of imposition on animals. The establishment of discrete levels or categories of imposition may be misleading, because stimuli and responses are each continua. Any categories are arbitrary, but they are set up to provide some reference points of biological significance.

Stimulation beyond the capacity for complete adaptation is a phenomenon which is referred to from here on as **stress**, a use similar to that in everyday speech. The first aim of this chapter is to investigate how the process of adaptation can be related in a logical and scientific way to the concept of stress. From this, the wider notion of **welfare** will be developed and the significance of the term **suffering** considered.

4.1 STRESS

4.1.1 Definitions

In common usage the term 'stress' implies exposure to unpleasant conditions with adverse effects. However, searching in popular writings for a description of what exactly qualifies as unpleasant and what is adverse does not produce simple answers.

From its earliest reported usage in 1440, according to *The Shorter Oxford English Dictionary*, stress has referred to 'the physical pressure exerted on an object', but also to 'the strain of a load or weight'. Stress has thus been used to refer to both the cause of a disturbance and its effect. Recent dictionaries perpetuate this dual meaning.

A sense of differentiation between the force causing an effect and the effect caused by it came from studies of the elasticity of materials by the physicist Robert Hooke (1635–1703). Hooke's findings, originally reported in Latin, are most widely referred to in the form 'strain is proportional to stress'. In physicists' terms, strain is the deformation produced when a body is subjected to a stress, and stress is the force producing that deformation. Stress is thought of as the causative agent, strain as the response. The physicists' differentiation between strain and stress has had limited influence on common use. In the current *Oxford English Dictionary*, the terms have virtually identical definitions.

The definitions of stress and strain originally proposed for physics have been used in various branches of science over many years. They have appeared in physiology (Cannon, 1935), ethology (McBride, 1980), animal science (Yousef, 1984; Moberg, 1987b), psychology (Lee, 1966; Fletcher, 1991) and ecology (Underwood, 1989). A return to these definitions in studies of 'stress physiology' has been advocated (Yousef, 1984; Moberg, 1987a), and this has been done in the standardized definitions for thermal physiology (Simon, 1987).

A major influence on efforts to standardize definitions of stress has been the work on biological adaptation to adverse environments by Hans Selye. Although aware of the original usage of stress and strain in physics (Selye, 1976), Selye established an alternative vocabulary in which he suggested that stress was the biological consequence of exposure to adverse environments (Selye, 1973). Selye described the adverse conditions themselves as 'stressors', and the processes of responding to stress as 'stress responses'. He also used the word stress in other ways, but these particular usages have been taken up by many people in medical and biological literature. The terms have been used in these ways despite their lack of precision or consistency with either common usage or terminology in the physical sciences.

Confusion arising from ambiguity in vocabulary is compounded by the uncertainty about what each term means (Dawkins, 1980). This uncertainty has reached such a level that some authors have concluded that we would be better off without the term 'stress' (Rushen, 1986c), have refused to define it (Becker, 1987), or have recommended that the term be avoided (Freeman, 1987).

Such problems in scientific communication are not easily solved. Even if abolishing the word were possible, the concept of stress is so important that its absence would prompt the coining of another word to replace it. Collective nouns, such as cancer and diarrhoea, as well as behavioural terms like aggression, bonding and fear, have great biological significance and serve useful functions, even if detail about their biology is incomplete. What is required in the case of 'stress' is a logical analysis of the biological basis of adaptation to adverse conditions, and the development of more strictly defined terms for scientific use which are not too different from the meaning ascribed to the term by the general public.

4.1.2 Selye's concept of stress

Before endeavouring to formulate a scheme relating stimulation, adaptation, stress, welfare and suffering, the contributions of Selye should be acknowledged and described. Selye's research drew attention to the fact that a wide range of adverse environments apparently evoked a limited range of responses. In particular, he emphasized that secretion of adrenal glucocorticoids is a widespread, non-specific response, as are suppression of the immune system, and the formation of gastro-intestinal ulcers (Selye, 1950, 1973). Furthermore, he noted similar patterns of physiological response in a range of animal species. These he summarized as: first an 'alarm' reaction, then a stage of physiological resistance to the disturbance, and, if this continues for long enough, a stage of exhaustion of the adaptive processes, leading to death.

The apparent general applicability of these concepts encouraged other studies, many of which confirmed that similar physiological responses could indeed be produced by a variety of different environmental conditions. The impression of a consistency in physiological responses to adversity led to wide acceptance of Selye's ideas. Unfortunately this reached a point where the significance of reports that did not conform to Selye's hypothesis was ignored. In recent years it has gradually become apparent that Selye's theory is not sufficiently precise to form a basis for theoretical arguments.

Three reservations have been noted about Selye's theory. First, the biological response to adversity is not as coherent as proposed, so his theory should not be taken as an assumption in experimental studies. Second, similar patterns of physiological responses can occur following both stressful and manifestly non-stressful stimuli. Third, Selye's inconsistent use of terms has increased rather than dispelled confusion.

Similar physiological responses can follow a variety of adverse stimuli, but the pattern is by no means uniform. Detailed analysis of the hormonal response, using assays that were not available to Selye, reveals that the physiological response is much more variable than Selye contended. Work by Mason's group (Mason, 1968; Mason *et al.*, 1968a-f; Mason 1971, 1975a, b) has shown that whilst cold conditions increased the activity of the adrenal cortex of rhesus monkeys, other unpleasant and sometimes life-threatening situations which the monkeys would avoid if they could, did not lead to this response. For example, adrenal cortex activity decreased rather than increased when there was a gradual increase in environmental temperature to a level which required considerable corrective action. Also, there was often no response to haemorrhage, close confinement which had lasted for some time, or to a diet which was entirely non-nutritive, during which the monkey was eating, but effectively fasting.

A further example of an adverse situation eliciting no adrenal cortex response is dehydration in sheep (Thornton, Parrott and Delaney, 1987). High environmental temperature, lack of food and lack of water are all situations in which it is not biologically adaptive to show an adrenal response which would tend to use up energy and increase body heat production. In dehydrated sheep an adrenal response either is not initiated or is suppressed by more pervasive responses.

The major criticism of Selye's theory by Mason (1971, 1975a,b), Moberg (1985, 1987a), Trumbull and Appley (1986) and others is that the neuroendocrinological and other biological responses to adversity are varied and stimulus dependent. It is clear that not only the neuroendocrinological but also the behavioural and immunological responses to noxious stimuli extend across considerable ranges. Recent studies reveal a complex interactive network of relationships among various parts of the brain and body involved in responses to adverse environmental conditions. Peptides, such as corticotrophin releasing hormone, ß-endorphin and others, are released and have a variety of different effects within the brain and body which vary with the levels of other peptides present at the time. In addition, it has been found that the adrenal cortex and the immune system have feedback effects on the production of peptides that are active in the brain. Effective opioid peptides include ß-endorphin, enkephalins and dynorphin, receptor sites for which exist in the brain and various other parts of the body, including on lymphocytes. It is clear that there is no single stress response, but rather a wide range of physiological and other changes which, although overlapping in some components, are usually quite specific to circumstances. The biological response to stress is considerably less uniform than the response proposed as the central tenet of Selye's theory.

A second problem with Selye's hypothesis has arisen following more accurate and extensive analyses of the adrenal hormones released in the resistive stage of adaptation. Glucocorticoids are released in response to situations that are not normally regarded as stressful, including courtship, copulation and hunting (Broom, 1988b). Species differences also exist, and certain stimuli that elicit an undoubted adrenal response in one species cause little or no effect in others (Freeman, 1987). As a result, single adrenal indices must be considered questionable indicators of stress in many circumstances because of the poor correlation with adverse effects, the specific effect of different stresses, and the wide individual variation (Moberg, 1987a).

Selye's inconsistent use of the word stress raises a third difficulty with his concept. Sometimes even within the same publication, Selye and those writing in support of his theories have used stress to indicate an environmental factor, the process by which such a factor affects an animal, and the long-term consequences of these environmental effects. Different words are needed for these different meanings. Use of the word 'stress' to describe long-lasting responses to environmental exposure is particularly distant from the physicists' use of the term.

The final problem with the terminology introduced by Selye is that it often differs from common usage. Before stress came to be equated with adrenocortical activity or was used to mean a disturbance of homeostasis, definitions and usage of stress and strain always implied adversity. Yet Selye (1973) has argued that 'stress is not something to be avoided ... we can enjoy it by learning more about its mechanism'. Such use of the term can only lead to confusion. This can be circumvented by discriminating between 'stress', which certainly should be avoided, and 'stimulation', which is an integral part of life and cannot be avoided. In fact, a minimum level of stimulation is desirable, as indicated in common usage by the implied value of 'a stimulating environment'.

4.1.3 Other concepts of stress

It has gradually become apparent since the experimental studies of stress that were encouraged by the early studies of Cannon and Selye that psychologically disturbing situations are important causes of emergency responses. Mason (1971) emphasized that expectations and fears were important causes of adrenal cortex responses for animals. In humans, Spielberger (1966, 1972) distinguished physical from psychological threats, and reported that while all people showed responses to physical threats, those people who were described as having 'high anxiety' were likely to show more responses to psychological threats. Attempts were then made to define stress in terms of those threats and their effects on regulatory abilities.

McGrath (1970) defined stress as an 'imbalance between environmental demand and response capability'. Lazarus and Folkman (1984) emphasized the importance of the perception of such an imbalance, by defining stress as 'a particular relationship between the person and the environment that is appraised by the person as taxing or exceeding his or her resources and endangering his or her well-being'. These definitions are difficult to interpret in biological terms. A definition referring to capacity to adapt is more relevant biologically. Indeed it is better in this respect than the definition of stress which has been used by some people, as 'anything which causes an adrenal cortex response'. The ability of the individual to cope with whatever challenge is received has often been included in discussions about stress in human psychological literature. For other animal species, however, individual variation in coping has been mostly ignored when stress has been considered. Some psychologists tend to overemphasize the importance of perception and coping, and forget about physical threats to the individual as sources of stress. This no doubt arises partly from different methods of study used on human and non-human subjects.

Hobfoll (1989) defined psychological stress as a reaction to the environment in which there is a threat of – or real – loss of resources, or a lack of gain following an investment of resources. The definition apparently includes any reaction to the environment, even the smallest one, and depends on the meaning of 'resources'; in this context Hobfoll defines them as 'anything of value'. Hence it is the reaction to the loss of something of value to the individual that Hobfoll considers to be stress. The definition indicates a little of the nature of the psychological stress, but it is not clear how it fits a more general concept of stress. Although it is important to emphasize the psychological components of control mechanisms and stress responses, too sharp a distinction between psychological and physical stress can be confusing. It is not always at all obvious when a physical effect becomes a psychological one. Indeed Dantzer *et al.* (1983) propose that it is the subjective experience of an aversive stimulus that leads to the physiological response.

A second problem with Hobfoll's definition is that there may be factors that are not valued by the individual but should be, because their lack will cause problems; painful treatment for a serious disease would be an example. A third problem is the ever-troublesome difficulty of measuring value. A final problem raised by the definition and considered later in this chapter is whether stress is an environmental effect or a reaction. Other ideas about the meaning of stress are referred to later in this chapter.

4.1.4 A general model of stress and strain

Recasting definitions of widely used concepts is bound to be disruptive and meet with resistance. It may nonetheless be justified. Selye's definition of stress is not that used in everyday language and is difficult to use in a logical way. Restructuring the concept of stress may help to remove the imprecision and inconsistency of present usage, so any disruption should be offset by some gain. Research on the fundamental biology of stress and welfare would benefit from a rationalization of its terminology.

Selye's influential theory has a good deal in common with the ideas developed in Chapters 2 and 3. They all hinge on the responses to, and the consequences of, potentially detrimental pressures applied to a regulatory system which maintains some degree of homeostasis. Yet the discrepancies between Selye's ideas and those outlined here require changes in the scientific definitions of stress and strain if they are to accord with common speech and the usage established in physics and elsewhere. There are features of Selye's concepts that are inherently confusing, such as the concept that stress can be beneficial, so these must be explicitly refuted. An unambiguous, usable definition of stress is required.

We contend that problems with the concept and vocabulary of stress are best solved by simplification, not complication. At times, other words have been used to describe adverse effects on individuals, for example 'eustress', 'overstress' and 'distress' (Ewbank, 1973). Such words do not overcome the problem. The general public and the scientific community have continued to use stress to describe a situation in which environmental conditions have adverse effects on an individual (Sassenrath, 1970; Hails, 1978; Warburton, 1979; Stephens, 1980; Gross and Siegel, 1981; MacLennan *et al.*, 1982; Dantzer and Mormède, 1981, 1983, 1985; Price, 1985b); the present proposals allow that to continue.

The use of the word stress to refer to 'any displacement from the optimum state' (Block, 1985) is of no real value, as any environmental influence on any individual would have to be included when using such a definition. If any perturbation of homeostasis were called stress, the word would have to be used every time that an individual showed a response whose effect is simply to return body state to within the normal range, for example after a brief exposure to the warm sun is followed by temperature regulatory responses; such a situation is a simple adaptive response (e.g. Fig. 4.2a). It corresponds to what Sanford *et al.* (1986) call 'physiological stress', but we contend that to use stress for a circumstance where the simplest of regulatory responses occurs is unnecessary and misleading. The question which arises, therefore, is – how great must the regulatory attempts or the consequences be before the word stress should be used?

Selye and many others since (such as Banks, 1982) have drawn the line between 'stress' and 'no stress' at the point where adrenal cortex activity occurs. However, there can be substantial fluctuations in adrenal cortex activity which are not responses to adversity, for example diurnal rhythms (El-Halawani *et al.*, 1973; Kalin, *et al.*, 1985b) and large variations during early development (Freeman, 1971). Since adrenal cortex responses occur when no adversity exists and do not always occur during adversity (Section

4.1.2), such a definition of stress would be unreliable. We support Fraser *et al.* (1975), Moberg (1987a) and others who emphasize that to imply that stress refers only to a single physiological phenomenon such as adrenocortical activity is scientifically restrictive and unwarranted.

Another more perceptive attempt to decide on a criterion for use of the term stress is Moberg's concept of the pre-pathological state (Moberg, 1985; 1987a,b). A rapid growth of interest in the link between stress and disease followed writings such as those of Engel (1967), Henry (1976) and Henry and Stephens (1977a). A description of the sequence of events occurring when psychosocial factors affect disease processes in humans (Kagan and Levi, 1974) is developed by Moberg (1985) into a 'model for response of animals to a stressful event' (Fig. 4.1).

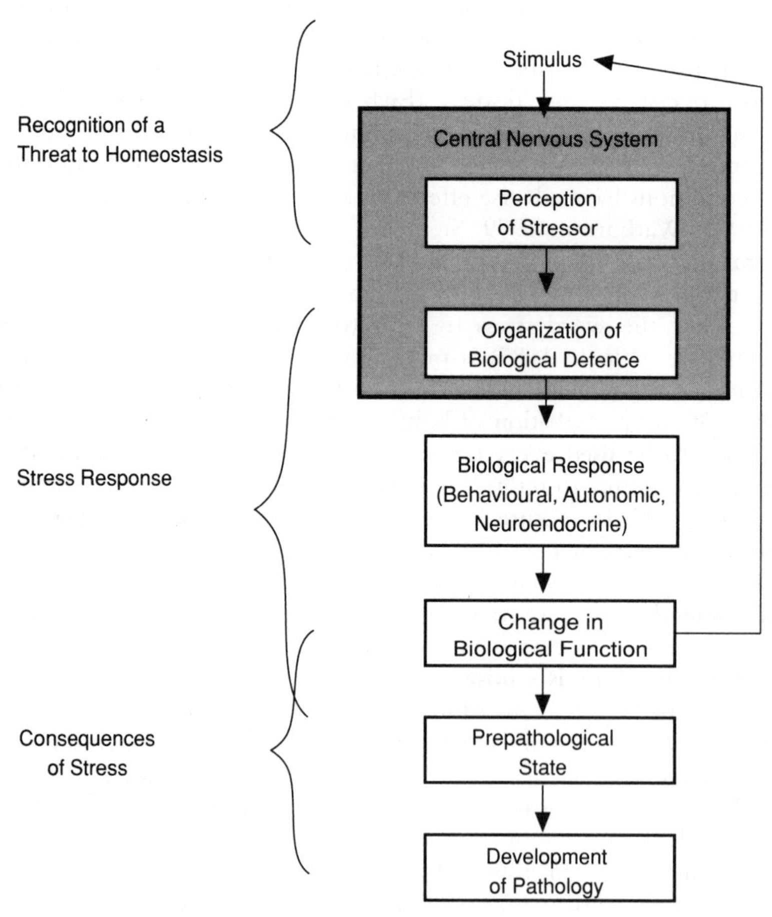

Figure 4.1 Model for response of animals to stressful event (from Moberg, 1985).

In this model, the crucial effect of the environment is to predispose the animal to the development of some pathological state. The pre-pathological state is recognizable by any of a variety of abnormalities and inadequacies in behaviour, physiology, immune system function and reproduction (Moberg 1987a,b). This concept of stress depends heavily on the meaning of the term 'pathology' which becomes crucial because some adverse consequences of the environment may not be considered pathological. However, the general concept outlined by Moberg is close to that developed here. One difference is that some pathological conditions, in which the pathogen has only a small effect which could not be hazardous to the individual, would be considered by Moberg to have some effect on welfare, but would not be considered a consequence of stress under our definitions.

(a) Coping

At this stage two concepts of great biological significance, **coping** and **fitness**, require discussion. As pointed out by Archer (1979) the widespread view that stress has adverse effects often includes some reference to a prolonged inability to cope with conditions. Coping at first glance seems so simple a concept as to warrant little attention, but this is deceptive. When an animal is imposed upon by a noxious stimulus, its biological response is either adequate to maintain stability and so enable the animal to cope, or it is inadequate, and the animal does not cope. In this sense, coping is no more than an indicator of the success the animal is having in controlling its internal environment.

This internal environment, however, is not just body physiology. As is evident from studies of human medical problems, many major difficulties in life are dealt with by mental control systems. Social interactions, sexual behaviour, carrying out the tasks which are necessary in daily work and aspects of basic body maintenance all involve complex mental control procedures which can fail to be effective. Other animals also have such control systems and such exposures to difficulty, even if the complexity of their world might be somewhat less than that of humans. The following definition takes account of all these aspects: to cope is to have control of mental and bodily stability (Fraser and Broom, 1990). Failure to have such control leads to reduced fitness.

It is clear from many studies that different individuals succeed to widely varying degrees in controlling their mental and bodily stability following a standard disturbance. Lazarus and Folkman (1984) emphasize the effect of such different capabilities in humans and, though less is known about other animals in this respect because it is harder to find out, there is little reason to doubt that they too may differ in their ability to adapt to a standard disturbing stimulus. Some individuals, and very likely some species, are better able to deal with difficulties, physical or psychological, than others; in other words, some cope better than others. There are individual and species differences in physiological as well as psychological responses to a given stimulus. This variation may be less of a problem in domestic animal welfare than it would be in the wild, because the controlled environment of domestic animals is often less variable, and the animals are bred for a specific purpose and so are more uniform genetically.

(b) Fitness

A second term, fitness, also needs introduction. Since it seems advisable to define stress by reference to its consequences, we are faced with the necessity of producing a precise criterion for what constitutes adverse or detrimental consequences. As we are dealing with a biological phenomenon, one possibility is to consider the fitness of the individual as the variable that is ultimately limited by exposure to stress. The idea that stress is something that reduces individual fitness has been put forward in a gradually evolving form by Broom (1983b, 1985, 1988c) and Fraser and Broom (1990). This concept provides arguably the most precise definition of stress. However, difficulties arise when applying it in practice.

The fitness of a genotype (a particular combination of genes) can be measured as the per capita rate of increase of the corresponding phenotype (the expression of those genes in an individual) (Sibly and Calow, 1983). As Sibly and Calow have pointed out, following Charlesworth (1980), the fitness of a phenotype in a particular environment depends on several basic life cycle variables: age at first breeding, interval between successive breedings, survivorship from birth to first breeding, survivorship of adults between successive breedings, and number of female offspring per female breeding attempt. The fitness of individuals subjected to particular environmental conditions can be calculated by measuring the variables listed above. If the environmental effect is to delay first or subsequent breeding, to increase the chance of mortality occurring before first or subsequent breeding, or to reduce the number of offspring produced, then fitness is reduced. The fitness of a genotype also depends on effects of the individual on others with that genotype but this will not be considered here as we are concerned with the fitness of an individual.

The effect of an environment on an individual is detrimental if the fitness of that individual is reduced, and in some cases that reduction can be clearly seen and can be measured (Chapters 5 and 6). In natural or near-natural conditions, reduction in fitness will usually be obvious, thus indicating the influence of stress and the existence of poor welfare. In many situations encountered in animal husbandry, however, it is not possible to be sure that fitness has been reduced, and it is necessary to use indicators that suggest that fitness is likely to be reduced. It is logical to imply that stress exists when there is a high probability that the fitness of an individual will be reduced, as well as when the reduction has been proven to have occurred.

(c) A definition of stress

Before presenting our definition of stress it is necessary to return to the question discussed at the beginning of the chapter of the confusing use of 'stress' to refer to a response as well as to an environmental factor. Perpetuating both the physicists' use of stress to describe an external force and the colloquial use of stress to mean the response of an individual is undesirable. When considering living organisms, there are problems with the strict physicists' definition of stress as a deforming force, because many environmental factors or forces exist without necessarily impinging on an organism. Hence it is best to define stress as an 'effect' so that it is quite clear that the organism is changed by some outside variable. The effect of stress occurring is to cause some or all of the control systems within the individual to work too hard for effective functioning, that is, to overtax them.

When considering the concept of stress, many people refer to loss of control, failure to cope and adverse effects on the individual. Figure 4.2

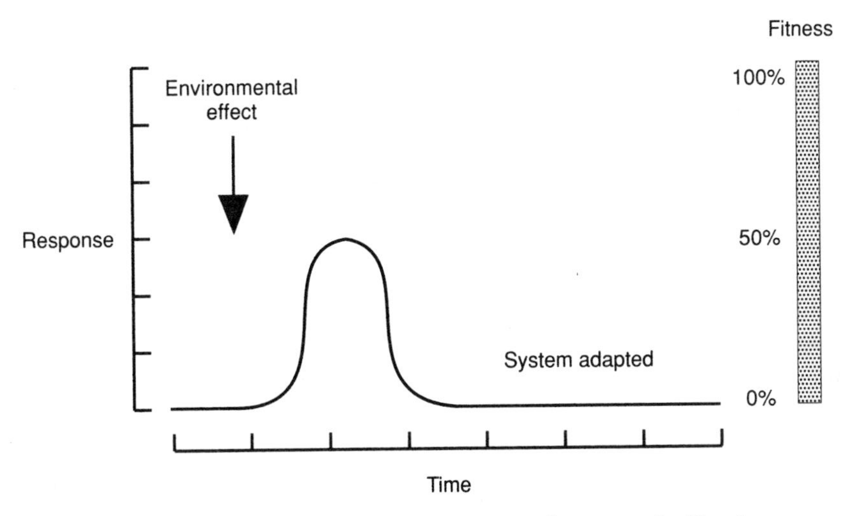

Figure 4.2(a) The schematic response to an environmental effect in a system which adapts fully; fitness is not reduced.

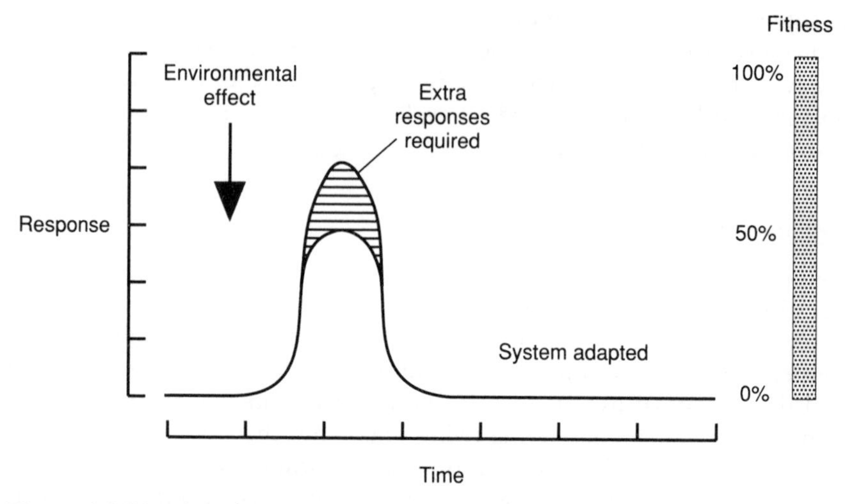

Figure 4.2(b) The schematic responses to an environmental effect which elicits extra responses in a system which adapts fully; fitness is not reduced.

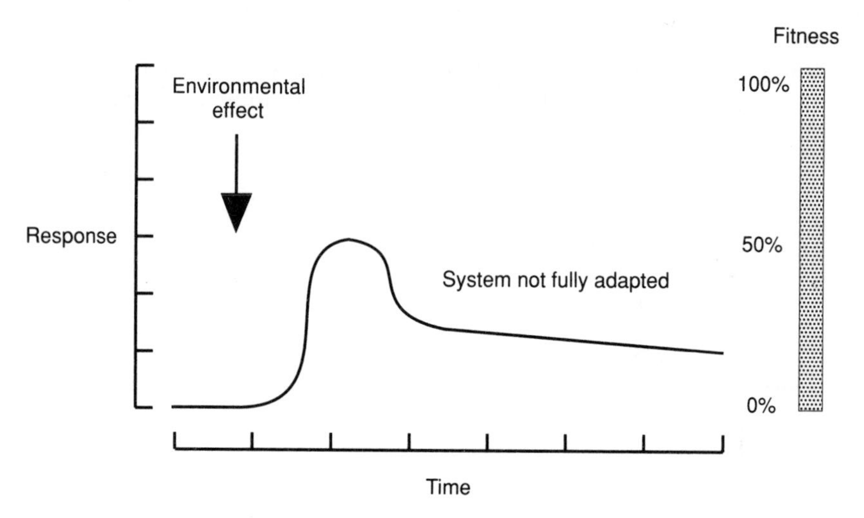

Figure 4.2(c) The schematic response to an environmental effect in a system which does not adapt fully; fitness is not reduced.

depicts the time courses of responses to various environmental effects and to their long-term consequences. Some responses have a simple regulatory effect, for example shivering or moving to a water source (Fig. 4.2a). Others, which are cross-hatched in the diagrams, are extra

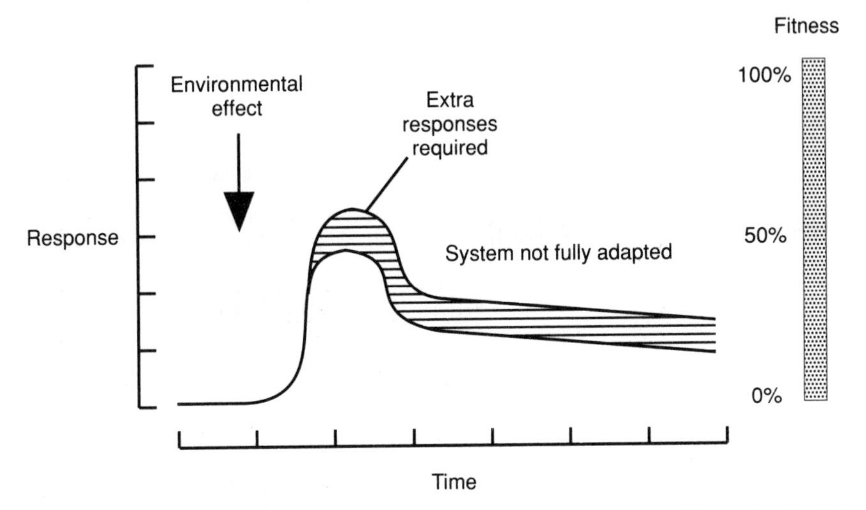

Figure 4.2(d) The schematic responses to an environmental effect which elicits extra responses in a system which does not adapt fully; fitness is not reduced.

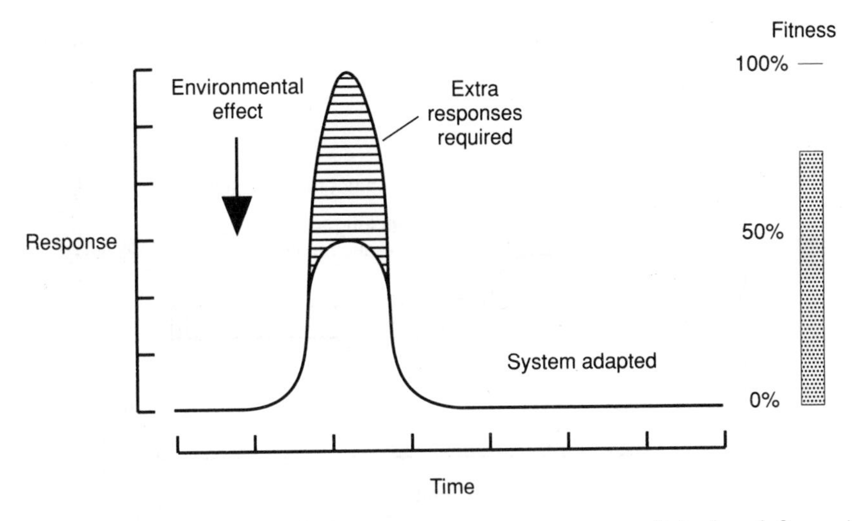

Figure 4.2(e) As Fig. 4.2b, but more extra responses are elicited and fitness is reduced.

responses which are utilized only when the simple responses are not likely to be sufficient for the body state to be maintained within the tolerable range (Fig. 4.2b). Such responses can include adrenal activity and various behavioural changes. The distinction between simple regulatory responses

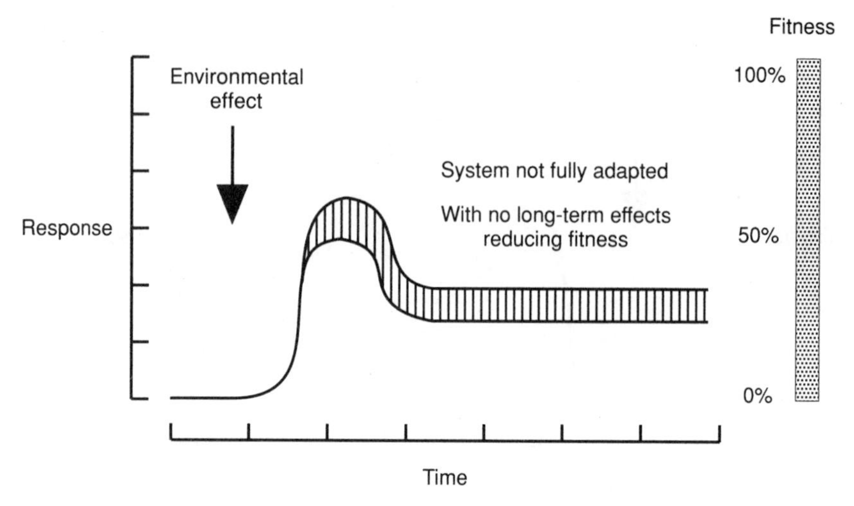

Figure 4.2(f) The schematic responses to an environmental effect in a system which never fully adapts; fitness is not reduced.

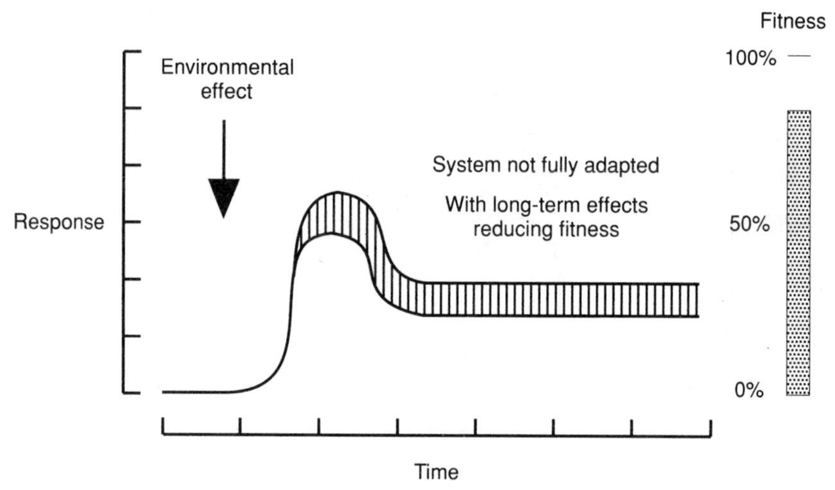

Figure 4.2(g) As fig. 4.2f, but fitness is reduced.

and extra responses which do not occur if the environmental effect is easily overcome and which may occur in a wide range of situations is useful, but not always clear cut. There may be different, specific regulatory responses which are activated after different amounts of effect on

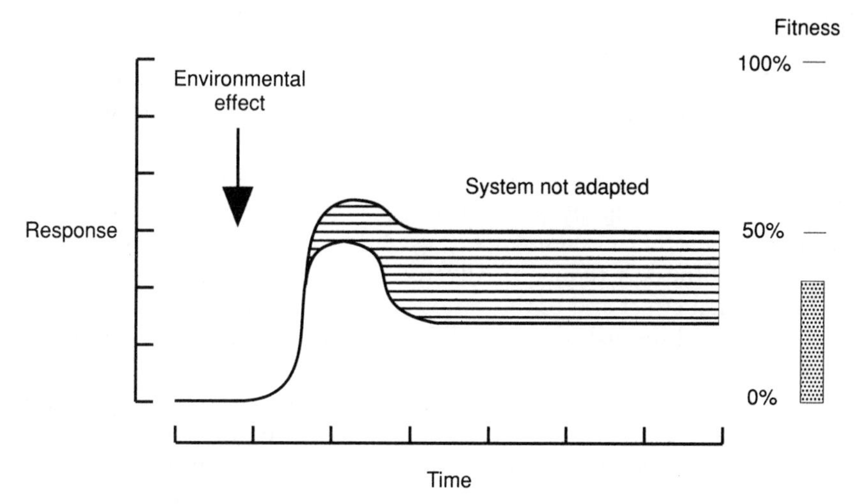

Figure 4.2(h) As Fig. 4.2f, but more responses elicited and more reduction in fitness.

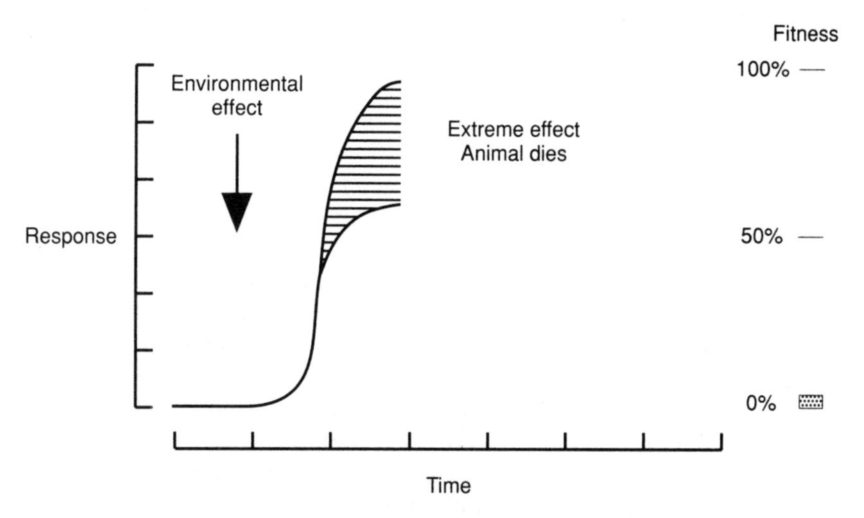

Figure 4.2(i) The schematic responses to an environmental effect which overwhelms the animal and drastically reduces its fitness.

the animal. Also, some behavioural responses, such as small birds mobbing a predator, may be secondary, but restricted to a narrow range of situations. Extra responses need not be associated with long-term adverse effects.

Responses may be required for only a short time (Fig. 4.2a, b, and e), or the system may not adapt for a longer time (Fig. 4.2c, d), or not at all (Fig. 4.2f–i). However, even if the continuation of a response is necessary for a long period (Fig. 4.2f), there may be no long-term effects of the kind which would reduce the fitness of the individual. Even if some energy were used up compensating for cool conditions or predator presence over a long period, survival and reproduction need not be affected. In another circumstance (Fig. 4.2g), the response level appears similar but fitness is reduced. Adverse long-term consequences of the environmental effect and of the various responses to it may be nil or they may be so substantial that the fitness of the individual is reduced (Fig. 4.2e, g, h, i).

As discussed above with reference to adrenal cortex responses, the differentiation between 'simple' and 'extra' responses is not precise enough to be used in the definition of stress. The continuation of a response to an environmental stimulus has been considered as a criterion for stress but, as mentioned earlier, some of the responses are so minimal that there is no adverse effect on the individual and the term 'stress' seems to overstate the severity of the situation when compared with common usage. However, if there is a reduction in the fitness of an individual, or if such a reduction seems likely to occur in the future, most people would consider the individual to be stressed, so this is the criterion for stress which is used here.

Therefore, we reach the conclusion that: stress is an environmental effect on an individual which overtaxes its control systems and reduces its fitness or appears likely to do so. There will normally be a reaction on the part of the individual to such an effect. This is a response to stress, or a stress response, and the immediate and short-term consequences of the stress are **strain**. The time course of the effect is not specified but, whether it lasts for a short period or for much of the animal's life, the animal is unable to cope with it. As explained in Section 4.1.4 (b), failure to cope implies reduced fitness. Some brief effects, such as being heated up a little by high environmental temperature, having an injection, or sustaining a minor injury, which are not likely to reduce the fitness of the individual would not be called stresses. Prolonged housing in boring conditions, a limited amount of immunosuppression, or long-term minor infection with a pathogen would be called stresses if they reduced fitness or appeared likely to do so but, on the other hand, could be sufficiently innocuous biologically not to constitute stresses. This definition of stress requires that there be an effect on fitness as well as an overtaxing of control systems (Fig. 3.6, level III). The definition also includes reference to circumstances in which the environmental effect appears likely to reduce fitness, though there may be no immediate measure of fitness. In practice there will be many situations in which it is not possible to be certain that the fitness of the individual will be reduced, but in which it is possible to deduce, using knowledge of the biology of the species, that this is likely. Such circumstances include substantial immunosuppression when disease challenge is likely, injury, behaviour abnormality, and physiological overload which increases the chances that food acquisition or the ability to avoid dangerous aggression will be reduced. A distinction is therefore made between minor disturbances to an animal's equilibrium, which may result in the use of energy to correct them but have no consequence for fitness, and those disturbances which do, or are likely to, reduce fitness. The differentiation between energy usage which does not reduce fitness and real fitness reduction which may, but need not, involve energy usage is important here and in other circumstances where biological efficiency and the way in which natural selection acts are being considered.

The term 'stress' is usually applied to members of the animal kingdom, but the arguments concerning its definition are also relevant to plants or micro-organisms. Hence, in contrast to 'welfare', which is generally restricted to animals, this definition of stress could be used for any living organism.

The state of the animal when it is stressed has often been called distress (Sanford *et al.*, 1986). Although stress and distress may well have been the same word originally (Stott, 1981), for the sake of clarity it is best to define distress less rigorously and to use it for a description of the state of individuals which are stressed or affected in similar ways by their environment. Reference is often made to pain and distress in order to include all aspects of the state of the individual.

4.2 WELFARE

When considering what animals do when they encounter problems that affect their normal functioning, it is important to distinguish those effects which reduce fitness or are likely to do so (i.e. stresses as defined above) from those which do not. There are many occasions when individuals find coping difficult, but succeed without long-term adverse consequences by, for example, using a brief adrenal response or a behavioural change of some kind. A minor injury or a period of illness might have no effect on the fitness of an individual. In each of these situations and on all occasions in which there is any kind of suffering, there is an effect on welfare even if there is no likely effect on individual fitness. Hence, stress invariably implies poor welfare, but welfare can be poor without stress, and welfare can range into the positive as well as being negative. The ideas developed in Chapter 3 (Figs. 3.4–3.7) are therefore developed further to include the terms welfare, stress and adaptation in Table 4.1.

Table 4.1 Gradations of stimulus and response in relation to usage of the terms adaptation, welfare and stress. It is important to emphasize that a small reduction in fitness may have less of an effect on welfare than very considerable and prolonged coping difficulties which do not affect fitness.

Level	*Environmental stimulus*	*Response by individual*	*Description*	*Welfare*
IV	extreme	instant death	lethal	very poor with death
III	noxious	coping attempts unsuccessful, fitness impaired	stressed	poor or very poor
II	aversive	coping with difficulty	adaptation, not stressed	poor or very poor
I	innocuous, with or without sensation	regulation carried out easily	adaptation, not stressed	unaffected

4.2.1 A definition of welfare

A definition of animal welfare is needed for scientific study, for legislation and for practical use. This definition must refer to a characteristic of an individual which is measurable. The measurement should be separate from any judgement about what is morally acceptable. Welfare, as a measurable characteristic of animals, should vary over a range rather than being something which exists or does not exist. Such variation is assumed in normal use of the term welfare, and is evident in its original use to indicate how well an individual 'fares' or travels through life.

As explained earlier, a wide range of problems is encountered during life and animals have a variety of methods for trying to solve or cope with these problems. Coping includes normal regulation of body state and emergency responses. The latter, e.g. high adrenal activity and heart rate, or flight activity, require more energy expenditure and hence are used only when the animal predicts that normal regulatory actions will be inadequate. The animal may succeed in its attempts to cope with the conditions in which it finds itself, in which case it has adapted to those conditions. Sometimes it may succeed only with great difficulty. Alternatively, it may fail to cope or seem likely to fail eventually to cope, and is stressed.

The welfare of an individual is its state as regards its attempts to cope with its environment (Broom, 1986b). The 'state as regards attempts to cope' refers to both how much has to be done in order to cope with the environment and the extent to which coping attempts are succeeding.

Attempts to cope include the functioning of body repair systems, immunological defences, emergency physiological responses and a variety of behavioural responses. An individual which is suffering is having more difficulty in attempting to cope than one which is not suffering and whilst it may be failing to cope, it need not be. Examples of indications of failure to cope are impaired life expectancy and reduced ability to reproduce. Hence, escape attempts resulting in a limited amount of pain, danger-induced adrenal activity for a limited period, or a short bout of depression may have no effect on life expectancy or reproduction, but cause poor welfare. If coping is difficult, or impossible, this will often be recognizable by scientific study of the individual. Our ability to measure coping difficulties, including those associated with pain and other suffering, and failure to cope, is improving.

This definition of welfare has several implications.

1. Welfare is a characteristic of an animal, not something given to it. In recent American usage, welfare can refer to a service or other resource given to an individual, but that is entirely different from this scientific usage. Human action may improve animal welfare, but an action or resource provided should not be referred to as welfare.

2. Welfare can vary between very poor and very good. One of the most important consequences of the arguments presented by Broom (1986b, 1988b, 1991a,b) and Fraser and Broom (1990) is that in order to use the concept of welfare in a scientific way it is necessary always to specify the level of an animal's welfare and not simply to reserve the word to indicate that the animal has, or does not have, problems. We should not speak simply of preserving or ensuring welfare, but of improving welfare or ensuring that welfare is good. We must be able to talk about an animal's welfare being poor when there is evidence that it is having difficulty in coping or is unable to cope. The continuum of welfare states is exemplified in Figs. 4.3–4.7.

3. Welfare can be measured in a scientific way that is independent of moral considerations. Welfare measurements should be based on a knowledge of the biology of the species and, in particular, on what is known of the methods used by animals to try to cope with difficulties and of signs that coping attempts are failing. The measurement and its interpretation should be objective. Once the welfare has been described, moral decisions can be taken. Ethical considerations are discussed in Chapter 8.

4. An animal's welfare is poor when it is having difficulty in coping or is failing to cope. Failure to cope implies fitness reduction and hence stress. However, there are many circumstances in which welfare is poor without there being any effect on biological fitness. This occurs if, for example, animals are in pain, they feel fear, or they have difficulty

controlling their interactions with their environment because of (a) frustration, (b) absence of some important stimulus, (c) insufficient stimulation, (d) overstimulation or (e) too much unpredictability (Wiepkema 1987; Chapter 3).

If two situations are compared, and individuals in one situation are in slight pain but those in the other situation are in severe pain, then welfare is poorer in the second situation even if the pain or its cause does not result in any long-term consequences, such as a reduction in fitness. Pain, or other effects listed above, may not affect growth, reproduction, pathology or life expectancy, but it does mean poor welfare.

5. Animals may use a variety of methods when trying to cope, and there are various consequences of failure to cope. Any one of a variety of measurements can therefore indicate that welfare is poor, and the fact that one measure, such as growth, is normal does not mean that welfare is good. This point is discussed by Broom (1986b, 1988b), Fraser and Broom (1990), Mendl (1991) and in Chapters 5 and 6.
6. Pain and suffering are important aspects of poor welfare. Pain is a sensation which is very aversive and suffering is an unpleasant subjective feeling, so it is also aversive and avoided where possible. Even though some pain and suffering may be tolerated in order that some important objective be attained, both of these involve increased difficulty in coping with the environment and hence poorer welfare. The relationship between welfare and suffering is considered again later in this chapter.
7. Welfare is affected by what freedoms are given to individuals and by the needs of individuals, but it is not necessary to refer to these concepts when specifying welfare. Needs and freedoms have been considered in Chapter 2 and their relationship to welfare is discussed further in Chapter 8.

4.2.2 The welfare concept and welfare measurement

In order that the definition of welfare is more easily understood and can be related to other concepts it is useful to consider examples of situations in which welfare is measured. Tables 4.2 and 4.3 list some measures of welfare. These measures are described in detail in Chapters 5, 6 and 7.

An animal encountering a varied environment but with no real problems will show only occasional bouts of adrenal cortex activity, and its welfare can be considered to be good (Fig. 4.3).

If that same individual was frequently frustrated or frightened, an appropriate indicator of this situation might be increased glucocorticoid production and synthetic enzyme activity in the adrenal cortex (Meunier-Salaun *et al.*, 1987). If high levels of adrenal cortex activity occurred very

Table 4.2 Measures of poor welfare

Measures of poor welfare
Reduced life expectancy
Reduced ability to grow or breed
Body damage
Disease
Immunosuppression
Physiological attempts to cope
Behavioural attempts to cope
Behaviour pathology
Self narcotization
Extent of behavioural aversion shown
Extent of suppression of normal behaviour
Extent to which normal physiological processes and anatomical development are prevented

Table 4.3 Measures of good welfare

Measures of good welfare
Variety of normal behaviours shown
Extent to which strongly preferred behaviours can be shown
Physiological indicators of pleasure
Behavioural indicators of pleasure

frequently or over a long period, a consequence may be widespread pathological changes. In this case the animal's welfare is even poorer. Sometimes the adverse effects may be so severe that adrenal function itself is impaired.

Other measures also allow the position of an individual on the welfare scale to be identified. Stereotypies are repeated relatively invariant sequences of movements which have no obvious function. Stereotypies such as bar biting or sham chewing in sows, tongue rolling in calves, crib biting in horses, or route tracing in zoo animals are often exhibited in conditions where an animal is frustrated or otherwise lacking in control over the world that impinges on it. The welfare states of individuals which show stereotypies for varying amounts of time are indicated in Fig. 4.4.

It is possible that a stereotypy that has been carried out for a long time relates more to an earlier state of an animal than to the present condition (Mason, 1991b). None the less, high levels of stereotypies, self-mutilation or other abnormal behaviour do show that an individual has difficulty coping with the conditions that exist at the time of observation, so the welfare of an animal that shows such behaviour is clearly poorer than that of an individual which does not. Even if the animal's problems were greater in the past and the abnormal behaviour has become a habit, the

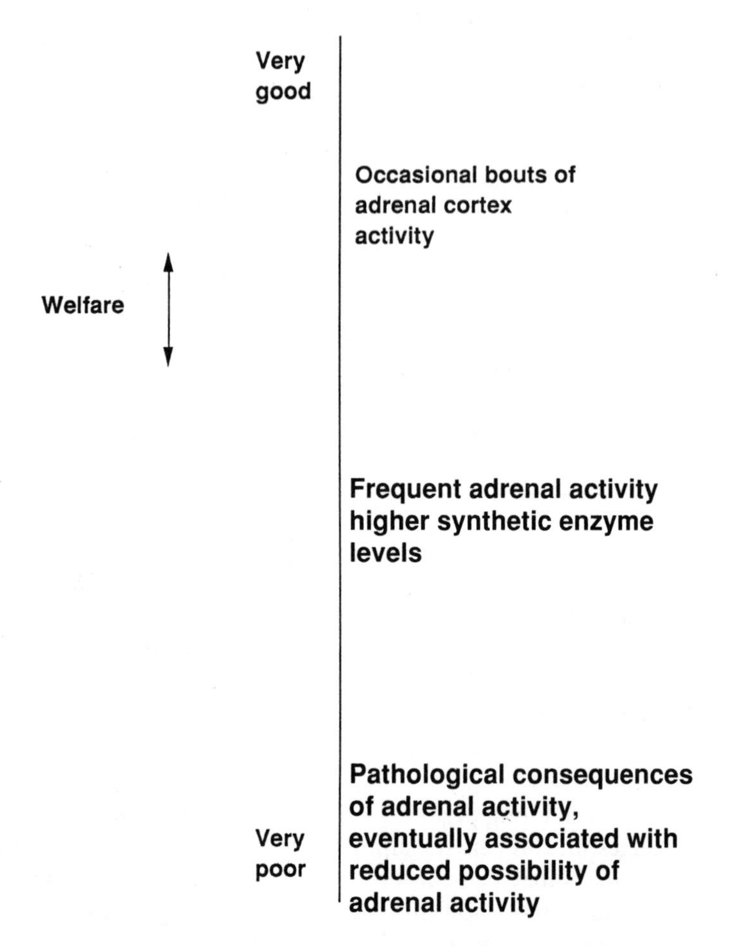

Figure 4.3 The significance for welfare of measurements of adrenal cortex activity effects.

behaviour will gradually, or rapidly, disappear if there is no current problem for the animal. A person who has been in solitary confinement in a prison camp may show some abnormalities of behaviour whilst recovering from that experience, and these indicate that the welfare of that person is poorer than that of a normal person. The more of life that is spent showing abnormal behaviour, the worse the welfare is.

A further behavioural response to conditions which pose problems for animals is to become inactive and unresponsive (Broom, 1987). Such unresponsiveness may be associated with increased influence of endogenous opioids, because it is linked with μ-receptor density in the cerebral cortex (Zanella, Broom and Hunter, 1991, 1992). This raises the possibility

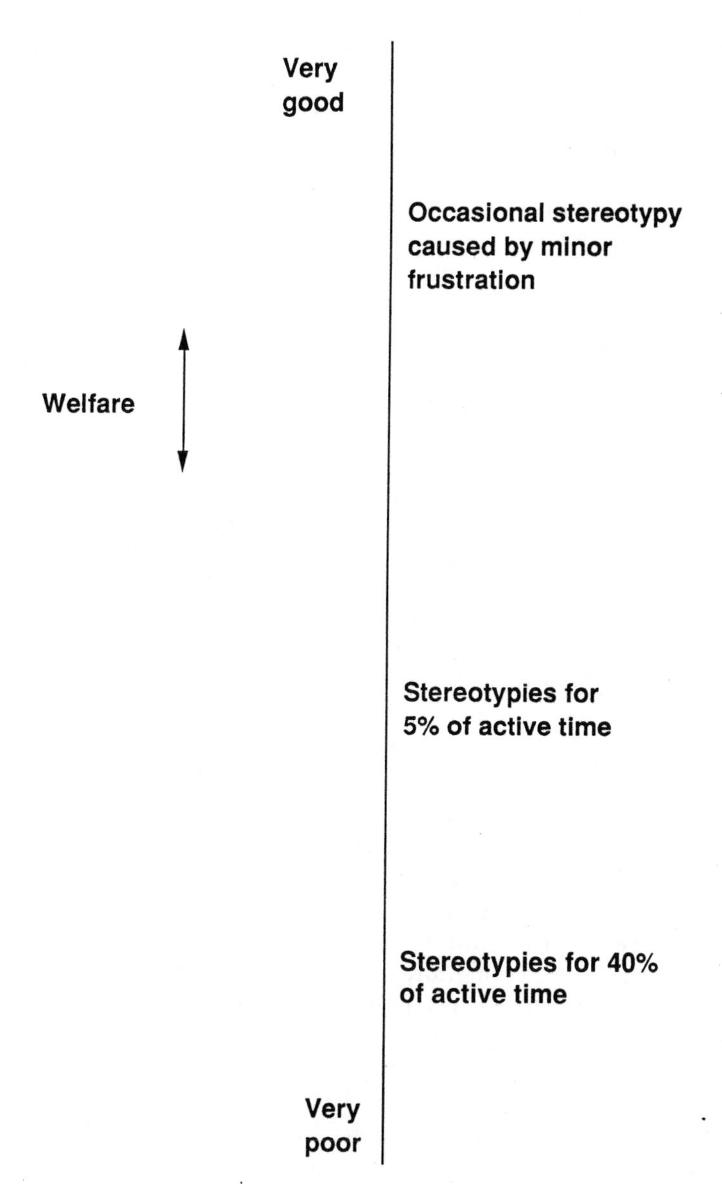

Figure 4.4 The significance for welfare of measurements of stereotypies.

of assessing welfare by measuring endogenous opioid levels. This would have to be done carefully taking account of all the roles of opioids.

Production measurements such as growth rate and reproductive output are also welfare indicators. For a given genotype, if growth or reproduction is impaired then welfare is poorer (Fig. 4.5) and again a scale of welfare based on measurements can be drawn up.

Other measures of welfare are discussed in the next two Sections and all measures are discussed in greater detail in Chapters 5 to 7.

4.2.3 Welfare in relation to suffering and other subjective feelings

The subjective feelings of an animal are a very important aspect of its welfare. Pleasant and unpleasant feelings are part of the experience of an individual as it attempts to cope with its environment. Dawkins (1980) has discussed suffering in some detail and has stated (Dawkins, 1990) that 'suffering occurs when unpleasant subjective feelings are acute or continue for a long time because the animal is unable to carry out the actions that would normally reduce risks to life and reproduction in those circumstances'. This description of suffering has clear links with the definition of welfare presented here.

Each of us knows when we have unpleasant subjective feelings. It is difficult to appreciate the subjective feelings of other people, indeed we can never be certain about them, but we accept that they exist and that they are similar in their general nature to our own. Our knowledge of the complexity of the organization of the brain and behaviour of other vertebrate animals is such that it now seems inconceivable that these animals do not also have subjective feelings. Indeed there are increased survival chances associated with having mental constructs which allow individuals to act in the absence of, or in advance of, specific stimulation and to change behaviour and physiological state in an adaptive way when there is some continuing lack of control over interactions with the environment. Such constructs are, essentially, feelings to which animals respond, and there must have been strong selection for those parts of the genotype which make such feelings possible (Chapter 7). It is also possible, however, that some feelings are an epiphenomenon of neural activity which do not confer any advantage on the animal. For instance, genes which promote the development of neural activity, and incidentally the feelings, might spread in the population because of their important primary effects, despite the fact that the feelings which neural activity produces are not useful or may be harmful. The general issue of the awareness and the feelings of animals has been reviewed by Griffin (1981), Dawkins (1990) and Duncan and Petherick (1991).

Although there can be little doubt about the existence of feelings, recognizing and assessing them is a considerable problem. There may be behavioural or physiological changes directly associated with some feelings, but none can be used reliably at present. Suffering is normally avoided if possible, so it may be clearly indicated by avoidance behaviour, especially if it involves much work and hence is quantifiable using operant techniques, may indicate it. However, suffering could obviously occur

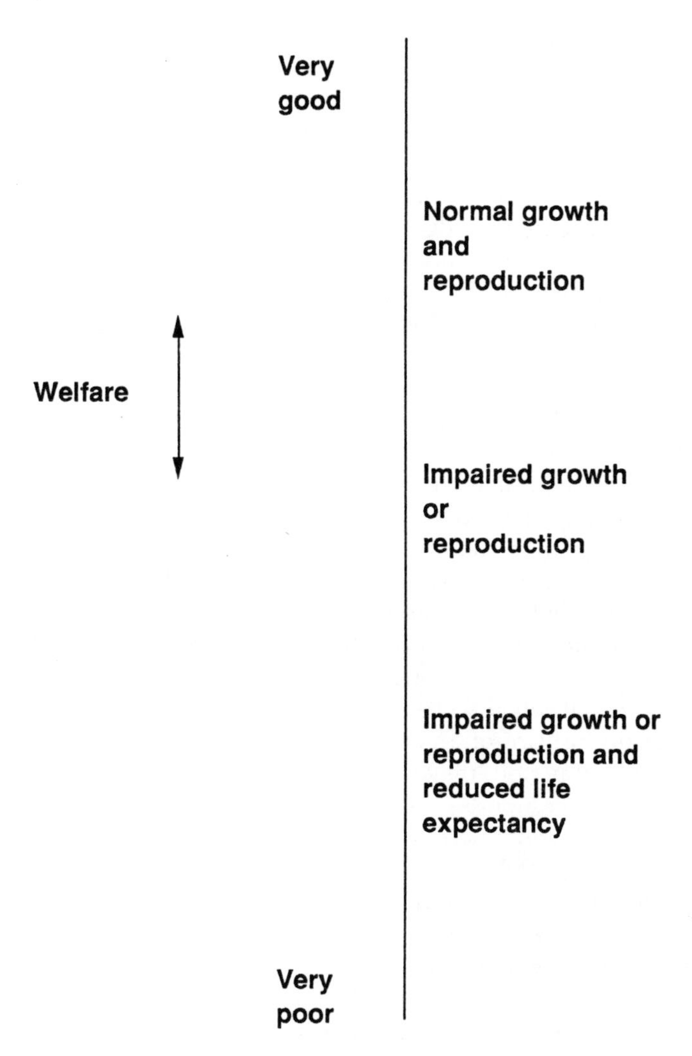

Figure 4.5 The significance for welfare of measurements of growth, reproduction and life expectancy.

without any avoidance being shown at the time. Situations in which suffering has previously occurred will usually be avoided if possible.

Dawkins' description of suffering, stated above, refers to feelings which are a consequence of inability 'to carry out actions that would normally reduce risks to life and reproduction'. Most people would include all but the milder, briefer kinds of pain and many of the consequences of disease within the term suffering. A definition of suffering should be somewhat broader than Dawkins' description. Not all pain and relevant disease effects are associated with the inability described, and suffering resulting from pain and disease does not necessarily result in risks to life or reproduction. Dawkins emphasizes one of the important situations in which suffering occurs, but a better definition might be: suffering is an unpleasant subjective feeling which is prolonged or severe.

Suffering is one of the most important aspects of poor welfare. We should all be concerned to identify suffering and to try to prevent it. If there is suffering, then the welfare will always be poor. However, welfare should not be defined solely in terms of subjective experiences, because situations arise in which the welfare of an individual can be affected without suffering occurring (Broom, 1991b; in press). For example, if an animal is injured, say by a bone breakage, a cut in the skin or an ulcer in the stomach, its welfare is poorer than that of an individual which is not injured (Fig. 4.6).

Even if the individual with the injury is asleep or anaesthetized, and hence not suffering, there is an effect on welfare. If there is suffering as well as an injury, then the welfare is poorer still. Few people would consider that a severe injury has no effect on welfare during sleep and that the welfare suddenly becomes poor on awakening because of consequent perception. It is difficult to see how the term welfare is to be used if there is extreme adherence to the concept (Duncan and Petherick, 1991) that only feelings count when welfare is being assessed. For example, should we consider the welfare of a person who is close to dying to be wholly good if, for a brief period, they experience a temporary good feeling?

Levels of immunosuppression also correlate with levels of welfare, as shown in Fig. 4.7, and described by Kelley (1980), Broom (1988b), and Fraser and Broom (1990). Levels may be mediated via hyperactivity of the adrenal cortex, but need not be. Changes may be preceded or accompanied by behavioural indicators of poor welfare. When welfare is good the immune system works effectively to counteract challenge by pathogens. There are sophisticated interrelations between the immune system and the brain. If there is immunosuppression, however, the animal will have to do more to cope with environmental challenges and will also show some of the pre-pathological effects discussed by Moberg (1987) which have the potential to reduce fitness; for both of these reasons the welfare is poorer than in an unaffected individual. It may be that the immunosuppressed individual does not suffer because it is not challenged by pathogens, but there is still an effect on its welfare, in that it is more vulnerable. If successful pathological attack occurs, with consequent morbidity and suffering, then the welfare is poorer still.

Suffering is an important concept when considering the effects of conditions and procedures on animals, but it is not necessary to try to equate it with poor welfare.

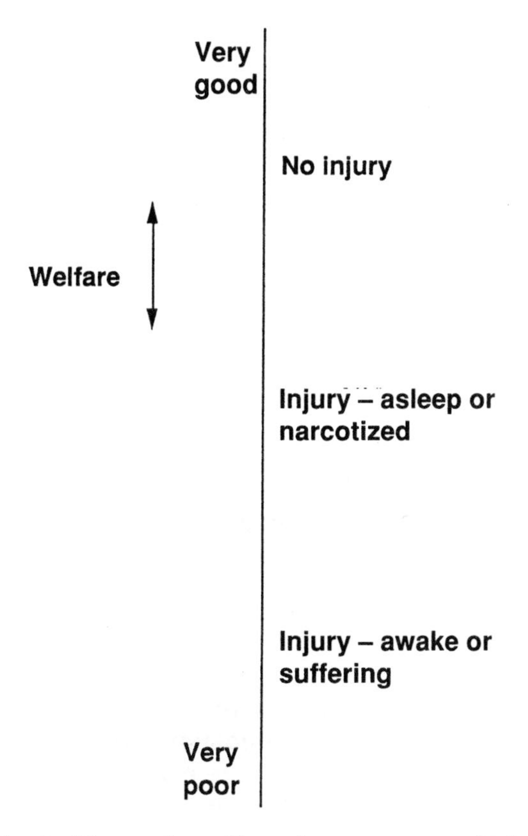

Figure 4.6 The significance for welfare of measurements of injury in relation to the extent to which the individual might be suffering.

4.2.4 Other behaviour measures and welfare

Some behaviour which is shown in response to adversity is clearly involved with attempts to cope with, for example, low temperature or lack of water. Other behaviour indicates how aversive a particular situation or stimulus is, and the degree of aversion can sometimes be assessed (Section 2.8.3). However some behaviour appears to be unadaptive and must be considered pathological, for example most forms of self-mutilation. This distinction between behaviour pathology and adaptive attempts to cope is often difficult to make, as mentioned at the beginning of Chapter 6.

Another measurement of behaviour listed in Table 4.2 as an indicator of poor welfare is suppression of normal behaviour. This measure assumes

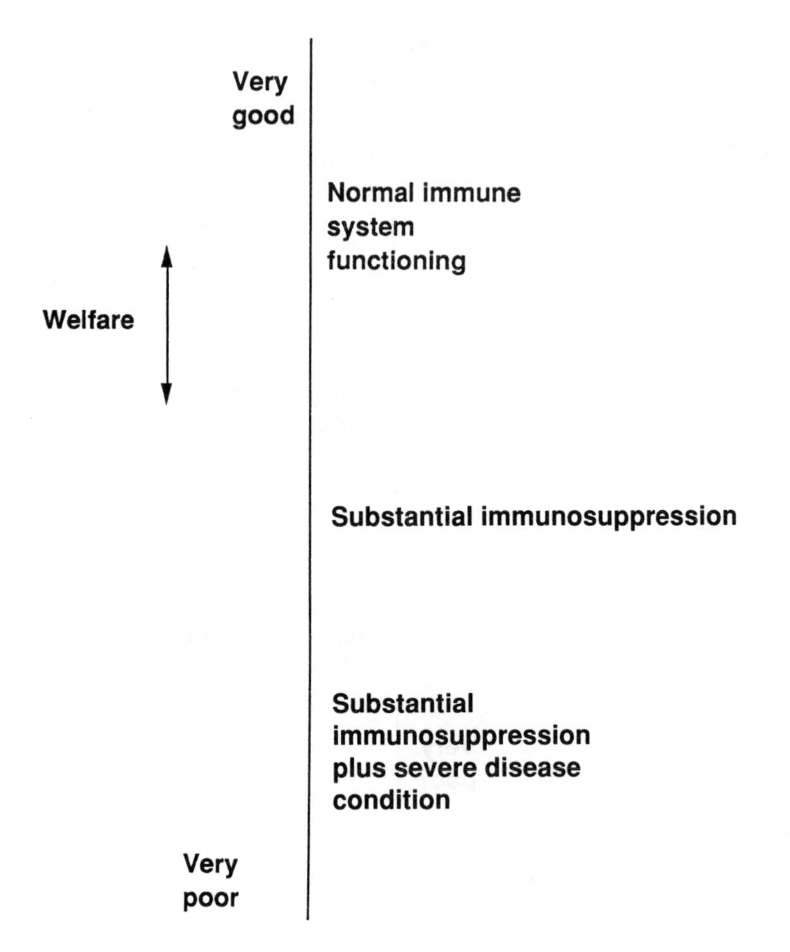

Figure 4.7 The significance for welfare of measurements of immune system function, disease condition and, hence, possible suffering.

a knowledge of normal behaviour in conditions in which the animal can allocate its time and energy as it chooses; such good conditions would be best identified in studies of captive animals, since conditions are often not optimal in the wild. If conditions are such that normal actions such as grooming behaviour, exercise or social interaction are not possible, then this will immediately be apparent when comparing animals in such conditions with those in good conditions, and it is reasonable to assume that their welfare will be less good. Failure to show normal anatomical development and physiological functioning may also be seen in such comparisons. Considerable knowledge of the species concerned is necessary in order to be sure that the absence of a particular behaviour or physiological processes is not just a consequence of, for example, lack of appropriate hormonal state. Such evidence of poor welfare is commonly reinforced by the presence of other indicators of poor welfare in the animals.

Good welfare is generally associated with a wide range of normal behaviour, shown in Table 4.3, especially behaviour which can be demonstrated to be strongly preferred (Chapter 7). If all preferred behaviours can be shown, then welfare will be better than if some are prevented. An ideal for animal welfare research is to recognize pleasure in individuals by means of physiological or behavioural measurements. However we cannot do this well at present. The difficulties of recognizing pleasure are illustrated by the following examples: the enormous number of self-stimulation sites found in the brain, since not all of these can indicate pleasure centres; the fact that people may misinform us about what is pleasurable, so verbal reports may be wrong; and the misinterpretation of behaviour such as tail wagging by dogs, which is often thought to indicate pleasure but may indicate submission.

.2.5 Welfare in relation to health

The term welfare, as defined here, is wide ranging in that it refers to the state of an individual as it attempts to cope with its environment. An important impact of the environment on an individual is that of pathogens or of pathology-inducing circumstances. This is the area to which the word 'health' is commonly applied, and health refers to the state of an individual. Welfare therefore includes health: indicators of good or poor health are also indicators of good or poor welfare.

.2.6 Linguistic problems with 'welfare' and 'well-being'

Some people use the words welfare and well-being interchangeably. The term well-being is not generally used in legislation or scientific writings because its meaning is less precise than that of welfare.

Welfare is the word used in English versions of modern European legislation. Most other languages have only one word to translate both welfare and well-being. The words equivalent to welfare in other languages, and used in comparable legislation include: *Wohlbefinden* in German, *welzijn* in Dutch, *bien-être* in French, *bienestar* in Spanish, *bem estar* in Portuguese, *welfaerd* in Danish and *dobrostan* in Polish. The common usages of some of these terms can make it difficult to use them to refer to poor, as well as good, welfare, though this is not a problem with *Wohlbefinden* or *welzijn*. However, it does seem surprising in colloquial usage to use a word meaning poor or bad to qualify *bien-être*, *bienestar* or *welfaerd*. It is generally desirable for a word in each language to be defined in a way that makes it equivalent to 'welfare' for scientific and legal use. Hence, as is necessary for all words used in a scientific or legal way, there should be international consistency of terminology, and the word for welfare in each language should be one which can be qualified with words for 'good' and 'poor'.

Chapter 5

Assessing welfare: short-term responses

Since our definition of welfare refers to the state of an animal, we should be able to use measurements of that state to indicate welfare. Many aspects of an individual's biology can reflect its attempts to cope with its environment, because there are various ways of trying to cope as well as numerous indicators of failure to cope. There can also be signs that welfare is good. It could be that, to combat some problem, one particular coping method is mainly employed, so measurements of that method would provide most of the necessary information. In most studies of welfare, however, it is desirable that a range of measures be obtained. Measures or techniques that are currently proving to be of value in assessing welfare are reviewed in this and the following two chapters. In practice, measurements of poor welfare are more common than those of good welfare, since poor welfare is associated with more obvious behavioural, physiological and pathological signs. Responses to, and consequences of, short-term problems are considered in this chapter.

Some methods of trying to cope with problems are used for both transient and long-lasting problems, but most methods are concerned principally with one or the other. Some indications that an individual is failing to cope may not be evident when a problem is brief because they occur only when the problem is long lasting; these are dealt with in Chapter 6. Short-term problems, the methods employed to cope with them and any other measurable effects which such problems have are the subject of this chapter. Short-term is taken here to mean lasting a few hours, whereas long-term problems are those lasting for a day or more. Short-term problems include many in which pain is felt, but anxiety, for example, often lasts long enough to be excluded from this chapter. Any distinction on the basis of time is somewhat arbitrary, especially because biological scaling suggests that small animals are affected more quickly than large animals, so differences in the pace of living must be considered when carrying out practical studies.

Situations that can lead to short-term welfare problems for animals include human intervention by close approach, handling, certain training methods, transport, some operations to help the animal, other operations carried out for convenience or vanity, deliberate cruelty, procedures preceding the slaughter of domestic animals, and some instances in which wild animals are trapped or killed. Other situations are not caused directly by human intervention; these include accidents, and attacks or threats by predators or animals of the same species (conspecifics).

The consequences of these situations may be obviously adverse for the individual, for example severe injuries, or they may be masked and become evident only through specific tests. Attempts to cope with the situations may be adaptive and beneficial to an animal, or may be harmful – for example, when caged animals chew their limbs or pluck out their own feathers. In either case they are indicators of poor welfare.

5.1 BEHAVIOURAL MEASURES

5.1.1 Orientation, startle and reflex responses

The most obvious indicator that an individual is experiencing difficulty in coping with a problem is often a behavioural response. The first behavioural responses to environmental change are orientation reactions (Sokolov, 1960). The individual turns so that bilaterally placed sense organs effectively locate and evaluate a directional sensory input. Alternatively the nose, or other olfactory organ in some animals such as insects, is raised into the air stream so that efficient olfactory recognition can occur. A set of physiological changes which alerts the animal and prepares it for action is brought into effect. Orientation reactions are common to many types and intensities of stimulation and are not themselves indicators that the animal is encountering a problem. However they may be followed by startle responses and defensive or flight reactions. They may also signal the onset of prolonged behavioural changes which are a direct consequence of the treatment experienced. All of these are welfare indicators.

Startle responses, comprising postural changes, jumps and vocalizations, are more than orientation reactions and their intensity is related to the extent to which the individual has been disturbed. The disturbing effect of a particular sensory input, for example that resulting from a loud noise, the odour of a predator, or the sight of a branch moving, depends upon the characteristics of the input, the context in which it occurs and the previous experience of it. Hence, the same input might elicit a substantial, or a minimal, startle response. The response includes cessation of previous activity, such as resting, feeding or grooming, followed by initiation of immobility, a posture that allows flight, defence, a jump or other sudden movement, and often the production of characteristic sounds. These startle responses vary in detail from one species to another. Some species freeze when startled, others flee; some vocalize, others do not.

When a domestic chick is startled by a local stimulus it stops its previous behaviour, orients to the stimulus and may give a brief call before freezing (Broom, 1968). The duration of visual fixation of the stimulus and of freezing can be measured, as can later reactions, for example, the intensity and frequency of loud 'cheep' calls and duration of bouts of calling. The response varies with the age of the chick and its previous experience (Broom 1969a,b). Small rodents such as mice and rats also freeze when startled, usually after orienting with the ears and nose in a position to receive auditory and olfactory information. An alternative startle response is to make a sudden movement, such as withdrawal of the body by a hole-dwelling fish, jumping by a frog, or a quick movement to remove the animal from a perceived threat and perhaps to make it appear less suitable as prey.

Startle responses are often followed by defensive or flight reactions. Indeed the dividing line between startle and active defence or flight is often unclear. However, the various components of the response may be quite distinct, as in the case of the squirrel which hears a sudden sound, orients, jumps, and then either runs away or prepares to fight according to whether the sound comes from a large predator or a rival. Flight from real or imaginary danger is usually easy to recognize. Defensive behaviour may be more difficult to identify, since it includes activities ranging from growling by dogs, to threatened or actual butting by cattle, and prolonged immobility which may make animals difficult to detect by, or appear dead to, a predator. Further examples of such behaviour can be found in Edmunds (1974) and Broom (1981b).

The intensity, duration and frequency of these responses can be measured as an index of disturbance. Part of the behavioural response is the cessation of normal behaviour, so the delay before such activities are resumed can also be a useful measure. When young domestic chicks are startled, the duration of freezing, number of loud cheep calls, and delay before ground pecking and preening resumed are greater in chicks reared with a familiar moving object than in those reared in a bare pen (Broom, 1969b). Each of these measures showed that chicks reared with a preferred moving object were less disturbed by a novel stimulus than were chicks which had not had that experience.

In social groups, social facilitatory effects can be of great importance in determining responses to disturbance. Chickens in a broiler house which are suddenly disturbed by a loud noise or a human action may show a flight response with alarm calls, which in turn promotes a similar response in other birds. The more the birds respond, the greater the stimulus to other birds. This positive feedback may result in mass hysteria, which can also occur in other social species, for example in crowds of people.

.1.2 Individual differences in behavioural responses

There is considerable diversity in the responses, both behavioural and physiological, which animals show when disturbed. In addition to differences amongst species in such responses, there are differences amongst individuals within a species. When rats, mice, tree shrews (*Tupaia*) or pigs are threatened or attacked by another individual of their own species, some individuals actively fight whilst others are largely passive and avoid interactions as much as possible (Koolhaas, Schuurmann and Fokkema, 1983; Benus, 1988; von Holst, 1986; Mendl, Zanella and Broom, 1992).

The duration of a freezing response and the extent of panic behaviour or increased adrenal activity can be assessed, but the possibility that individuals may not all respond in the same way must be considered. Indeed, individuals may vary in the extent to which they show active and passive responses. In the study of chick startle responses mentioned above (Broom, 1969a,b), the first response was usually freezing, but this was replaced after some seconds or minutes by active attempts to escape. Hence these chicks were responding both passively and actively, and a simple measurement of activity could give a misleading assessment of response magnitude.

Variation in behavioural responses is also evident in pigs subjected to unfamiliar handling, driving and attempts to move them up a ramp on to a vehicle. Some pigs run away screaming, others freeze and are reluctant to move even when pushed, and a few pigs try to bite the person when approached closely. These are all responses to a situation which, for a pig which has not had much previous contact with people, is presumably a disturbing experience. Likewise a dog which is treated by a veterinary surgeon may show escape attempts or defensive snarling and biting. However, another dog may show withdrawal and reduced activity. Escape, active defence and withdrawal are all indicators of disturbance, the intensity and duration of which can be measured. In a study of the responses of cats confined in a cage in a veterinary hospital or animal shelter, McCune (1992) found that the most extreme responses were a retreat to the back of the cage, crouching and marked inactivity.

5.1.3 Behavioural indicators of pain

Behavioural responses are of particular value when we attempt to assess the extent of pain. Short-term pain elicits a substantial behavioural response in some species but very little in others. Characteristic responses include changes in posture during abdominal pain, avoiding use of a painful limb, and licking the painful region. Other responses to localized pain are tightly closed eyelids in cases of eye pain, holding the head on one side or shaking the head in cases of ear pain, rubbing the mouth and unwillingness to eat in response to mouth pain, and vigorous responses when any painful area is touched. Descriptions of responses to pain are to be found in Morton and Griffiths (1985), Flecknell (1985a,b), Zimmermann (1985), Duncan and Molony (1986), Robin (1986), Spinelli and Markowitz (1987), Wallace *et al.*, (1990), and Wright and Woodson (1990). With such descriptions, good measures of the extent of pain can be made.

The use of behavioural measures to assess the extent of pain is illustrated by the study of castration of piglets by Wemelsfelder and van Putten (1985). Piglets were castrated without anaesthetic during the fourth week of age, as occurs on many farms in the Netherlands. Handling itself elicits struggling and loud squealing from piglets, so the movements and vocalizations of female piglets which were handled and male piglets which were both handled and castrated were compared. The mean frequency of the scream during handling only was 3500 Hz but after the first cut during castration it was 4500 Hz and after the second cut it was 4857 Hz. Both the number of frequencies occurring in the sound and the number of changes in sound distribution over the frequency range were higher after castration. Recently castrated piglets were less active and showed more trembling, leg shaking, sliding and tail jerking; some vomited, and all initially avoided lying, then later lay in a way that appeared to spare their hindquarters. The duration of the discomfort was indicated by the continuation of some of these changes in behaviour for 2–3 days.

When there are behavioural responses which, it is suspected, indicate that the individual is in pain, a check is to administer an analgesic and observe any changes in response. For example, reduced levels of food and water intake and of activity following surgery in rodents were restored to near normal by analgesic (Flecknell and Liles, 1991; Flecknell *et al.*, 1991). There can be problems in interpreting experiments involving the use of analgesics because substances which are analgesics in one species may not be effective in another. Also, the analgesic may itself affect behaviour, for example opioid analgesics often increase activity in rats. Positive results, like those of Flecknell and Liles, which refer to diminution by analgesics of the effects of a potentially painful procedure on several different measures are evidence for the occurrence of pain, but lack of effect of a supposed analgesic provides no certain evidence for the absence of pain.

Further uses of behaviour in the assessment of pain are the tail pinch or tail warming experiments which give information about drug effects (Section 6.8) and the measurement of subsequent aversion after experience of a potentially painful treatment. Cooper and Vierck (1986) trained monkeys to pull a bar in order to escape electrical stimulation. By titrating willingness to pull the bar against the electrical stimulus, pain tolerance levels were found to be similar to those of humans. In a similar way, Zimmermann (1985) trained cats to terminate a painful stimulus, but noted that the cats would tolerate a certain amount of pain in order to obtain a food reward. In a similar experiment, Cabanac and Johnson (1983) found that rats would voluntarily go through a cold maze, which may have resulted in some pain, to obtain a preferred food even when normal food was available in warm conditions. The study by Rushen (1986b) of the aversion of sheep to returning to a place where they had experienced various handling treatments is described in Section 2.8.3. A similar kind of aversion was found by Fell and Shutt (1989) when they observed sheep subjected to the mulesing operation in Australia. This involves cutting away about 50 cm^2 of skin from the tail and genital region in order to reduce the risk of attacks by flies which lay their eggs in damp wool causing wounds and death. It is carried out without anaesthetic. There was little immediate behavioural response to the treatment, but mulesed sheep subsequently showed abnormal posture and locomotion, and strong avoidance of the people who had restrained them during the operation (Section 5.2.3).

5.2 PHYSIOLOGICAL MEASURES

5.2.1 Heart rate

Except in diving animals, increases in heart rate (tachycardia) occur when the level of physical activity of an animal and, hence, its metabolic rate, increases. But heart rate can increase before an action occurs, or it can decrease (bradycardia) as an emotional response to a situation, in humans even slowing to the point where the subject faints (Guyton, 1991). In some species such a response may be adaptive in the presence of a dangerous predator or conspecific, which may refrain from attacking an individual that appears dead or at least is not showing behaviour eliciting pursuit. Changes in heart rate prior to changes in behaviour have been described for domestic chicks by Forrester (1979) and Potter (1987). In general, the sympathetic nervous system causes the increase in heart rate and the parasympathetic nervous system the decrease.

Measurement of heart rate can be a useful measure of the emotional response of an individual to short-term problems, provided that distinction is made between the metabolic and emotional effects, and that the measurement itself does not cause too much disturbance. The system used to monitor the heart rate must not itself have an effect on the animal. Early work on free-moving animals required leads to connect the animal to the recording apparatus and, although some useful results were obtained, telemetric methods or recorders carried by the animal have proved to be better. Bohus (1974) and Adams *et al.* (1988) found that basal heart rates of rats and baboons were lower if a telemetric system was used than if leads from the animals were connected to apparatus outside the cage.

Bradycardia often occurs during the orienting reaction just after a stimulus is detected, but the major part of the response of most species is tachycardia. Exceptions are those species which show freezing responses. Gabrielsen, Kanwisher and Steen (1977) recorded the heart rate of incubating willow grouse (*Lagopus lagopus*) and found that there was bradycardia when a predator approached the well-camouflaged bird as it sat immobile on its nest. Bradycardia has also been reported in wild rodents during immobility (Hofer, 1970) and it occurred when laboratory rats froze in response to a drop in background noise level (Steenbergen *et al.*, 1989). When male tree shrews were defeated by other males in fights and remained passive thereafter, they exhibited bradycardia, in contrast to active fighters which showed tachycardia (von Holst, 1986). Bradycardia is also well-known as an adaptive response during dives by air-breathing animals which are well adapted for diving. Care must be taken to consider the biology of the animal when using heart rate changes as an indicator of welfare.

In farm animals, such as sheep (Syme and Elphick, 1982), heart rate changes according to their handling or other treatment by people. In a study by Stephens and Toner (1975), the heart rate of a calf standing quietly was 90 beats per minute, but this increased to 135 beats per minute when a person entered its pen and to 145 when it was restrained. Similarly, van Putten and Elshof (1978) found that the basal heart rate of 138 beats per minute in pigs was increased by a factor of 1.5 when an electric prodder was used on them and by 1.7 when they were made to climb a ramp. Telemetric studies by Duncan and Filshie (1979) showed that hens in cages displayed tachycardia when a person approached their cage. The response lasted longer in birds of a strain which manifested little behavioural response and was regarded as docile, than in birds of a strain which showed a considerable behavioural response. Heart rate also increased in dogs or rats which were expecting an electric shock (Liang *et al.*, 1979; Gomez, Büttner and Cannota, 1989).

A common problem in studies of heart rate is that changes as a result of metabolic activity cannot be distinguished from changes due to emotional responses. Baldock, Sibly and Penning (1988) and Baldock and Sibly (1990) overcame this problem by recording basal levels of heart rate of sheep engaged in normal levels of activity: lying, ruminating, standing, and walking (Table 5.1), and taking account of the level of activity.

These sheep, which had frequent human contact, showed no heart rate change when they were spatially isolated or put into a stationary trailer, but other treatments caused varying degrees of tachycardia, which was maximal when the animal was approached by a strange person with a dog.

Cardiac arrhythmias are sometimes a response to difficult situations. Liang *et al.* (1979) found that dogs anticipating an electric shock showed greater ventricular excitability. A variety of unpleasant stimuli resulted in arrhythmias in monkeys and rodents (Hofer, 1970; Corley *et al.*, 1973; Nyakas, Prins and Bohus, 1990). Cardiovascular measurements such as blood pressure are also influenced by stressful stimuli, but are more indicative of long-term than short-term problems.

Table 5.1 Sheep heart rate responses (bpm) (from Baldock and Sibly, 1990)

Treatment	*Change in heart rate (taking account of activity)*
Spatial isolation	0
Standing in stationary trailer	0
Visual isolation	+20
Introduction to new flock (0–30 min)	+30
Introduction to new flock (30–120 min)	+14
Transport	+14
Approach of man	+45
Approach of man with dog	+79

5.2.2 Respiratory rate and body temperature

Other physiological changes which can be affected in similar ways to heart rate are respiratory rate and body temperature. Increases in activity which cause tachycardia usually affect both these variables, and they can be easier to measure without disturbing the animal. The respiratory rate can be assessed by observation of a stationary animal from some distance, and is frequently monitored by veterinary surgeons treating animals. Changes in respiratory rate can occur during emotional disturbance without body activity (Mellor and Murray, 1989). In an experimental study by Broom *et al.* (in preparation), respiratory rate was measured in hens from a commercial battery house. The hens were removed from their cage and carried using the normal method of being taken to slaughter, held by the legs upside down with 2–4 other birds. Fifty seconds after they were removed from the cage the respiratory rate of birds restrained lying on their sides was 34 breaths per minute, and 2 minutes afterwards, while still restrained, it was 38 breaths per minute. As with heart rate, an increase in breathing rate could be a response to a situation perceived by the individual or it could merely reflect greater activity. The former is of greater interest as a measure of welfare.

Core body temperature fluctuates diurnally but can increase following disturbing events. Georgiev (1978) found that the body temperature of laboratory rats increased by a mean of 1.4 °C during a storm and also increased when an unfamiliar human attendant was present. Reite *et al.* (1981) found that infant macaque monkeys had elevated body temperature when they were protesting following separation from their mothers. However, the body temperature was reduced later when they reached an apparent 'despair' phase. Similarly, von Holst (1986) found a body temperature reduction in tree shrews which had been defeated by another individual. Again it is necessary to understand the biology of the animal and to identify the type of response which is being shown when trying to assess welfare by the use of this measure.

Handling and transport which causes increased adrenal cortex activity can also elevate body temperature. Trunkfield *et al.* (1991) recorded rectal temperature of calves after removal from normal housing, loading onto a vehicle, a one-hour journey and pre-slaughter handling. There was a significant increase in body temperature in crate-reared calves, which also showed the largest cortisol response to handling and transport. Although disturbing events can cause an increase in core temperature, peripheral temperature may drop in circumstances in which sympathetic nervous system activity causes vasoconstriction. For example, Duncan and Filshie (1979) reported a decrease in the external temperature of the lower leg in chickens following exposure to alarming visual stimuli.

.2.3 The adrenal axes

Measurements of activity in the sympathetic-adrenal medullary system and in the hypothalamic-pituitary-adrenal cortex system are amongst the most useful in the assessment of how difficult it is for animals to cope with short-term problems. As emphasized at the end of this chapter, there is considerable individual variation in how these and other coping methods are used, but changes in the levels of hormones from these two systems are frequently seen as responses to problems arising from environmental conditions, especially those which are severe for short periods. Both systems cause changes in the body which alter the range of substrates available for emergency action: more glucose after adrenomedullary hormones, more amino acids and fatty acids after cortisol. These have the effect of making energy for emergency action more readily available. They differ in their time course in that adrenal medulla hormones are shorter lived than adrenal cortex hormones, and adrenal cortex activity has more long-term effects. Both systems can be activated in beneficial and detrimental circumstances (see p. 61), so care must be taken to consider the context of their activation before deducing that any adverse effect on welfare has occurred. If it is clear that the behaviour being carried out is beneficial but involves increased activity, the possibility that some or all of any increase in adverse activity is associated with this activity must always be considered.

The principal products of the adrenal medullary response to emergency situations are the catecholamines adrenaline (epinephrine) and noradrenaline (norepinephrine). In humans, emotional disturbances of the kind which elicit a more passive response cause a greater increase in adrenaline production, and those disturbances which are associated with physical activity, particularly aggression, cause a greater increase in noradrenaline production (Mason, 1968; Frankenhaeuser, 1975; Goldstein, 1987). The release of these catecholamines from the adrenal medulla occurs within 1 or 2 seconds of the perception of the initiating stimulus, but their metabolism is very rapid, the half-life in rat blood being 70 seconds (McCarty, 1983). The amount of adrenaline and noradrenaline released into the bloodstream when environmental conditions which cause problems are encountered is related to the extent of the problem, so sampling of blood must be very rapid if any useful information is to be obtained. Only samples taken using an intravascular cannula are of use. Samples taken after killing cannot be used since Popper, Chiueh and Kopin (1977) demonstrated a 10-fold increase in plasma adrenaline and an 80-fold increase in plasma noradrenaline following decapitation of rats. Both of these increases are higher than those caused by experimental manipulations. Levels of adrenaline and noradrenaline in urine can provide some information but they are very variable (Baum, Grunberg and Singer, 1982).

When Kvetnansky *et al.* (1978) monitored plasma catecholamine levels in catheterized rats, they found that opening the cage door caused an increase, handling or transfer to a new cage caused a larger increase, and taping the animal to a board caused the greatest increase which was 40-fold for adrenaline and 6-fold for noradrenaline. Results which were similar, at least in part, were obtained by Popper, Chiueh and Kopin (1977), de Turck and Vogel (1980), Livesey, Miller and Vogel (1985) and Castagné *et al.* (1987). Individuals varied in basal and post-restraint values. Levels of catecholamine responses to foot-shock in rats were found to vary over a wide range, and in proportion to the shock intensity (Natelson *et al.*, 1981). Konarska, Stewart and McCarty (1989b) found increased plasma catecholamine responses to three types of experience (in ascending order): forced swimming in water at 18 °C, electric shock of 1×10^{-6} A every 5 seconds, and restraint in a plexiglas tube. Social interactions also result in increases in plasma catecholamine levels, as shown for example in Sachser's work (1987, Sachser and Lick, 1989) on male guinea pigs, especially those defeated in fights. The use of plasma adrenaline and noradrenaline as a measure of welfare in conditions lasting for a short period is of value, but should be limited to use in catheterized animals from which samples can be taken within one minute of treatment.

The first stage of activity in the hypothalamic-pituitary-adrenal cortex system is interleukin 1ß stimulated secretion of corticotrophin releasing factor or hormone (CRF or CRH). Measurements of CRF in the hypothalamus are possible but these can be made only in very restricted experimental conditions. The release of adrenocorticotrophic hormone (ACTH) from the adenohypophysis (anterior pituitary) is initiated principally by CRF although it can also occur in response to catecholamines

(Axelrod, 1984) or to the neurohypophysial hormones arginine-vasopressin and oxytocin (Gibbs, 1986b; Gaillard and Al-Damluji, 1987). ACTH is carried in the blood to the adrenal cortex where the glucocorticoids – cortisol, corticosterone, or both – are released. The production of ACTH and CRF is inhibited by glucocorticoids, and ACTH is removed quickly from the blood. Hence measurement of plasma ACTH levels must be made within a few minutes of the event whose effect on welfare is being assessed. Meyerhoff *et al.* (1988) measured both ACTH and cortisol in soldiers undergoing an oral examination, and found that ACTH levels were elevated earlier than those of cortisol and that they returned more rapidly to basal values. In the guinea pig, significantly increased plasma ACTH was measured 4 minutes after exposure to noise and vibration (Bailey, Stephens and Delaney, 1986). The peak ACTH response of rhesus monkeys to confinement and exposure to a loud noise occurred 15 minutes after onset, but the response then declined despite the continuation of the conditions (Kalin *et al.*, 1985a). Rats show progressively increasing levels of plasma ACTH in response to the noise of an alarm bell, restriction in a plastic tube and immobilization by being tied to a board (Armario and Jolin, 1989). Other treatments which produced ACTH elevation were electric shock (Rossier *et al.*, 1977; Onaka, Hamamura and Yagi, 1986; Maier *et al.*, 1986), a novel environment (Hennessy *et al.*, 1979; Gibbs, 1986a) and low temperature (Gibbs, 1984).

Glucocorticoid levels rise in response to many short-term problems in life and their measurement gives valuable information about the welfare of animals. Prior to the development of radioimmunoassays (RIA) and enzyme-linked immunosorbent assays (ELISA) for glucocorticoids in blood plasma or saliva, a metabolite 17-hydroxycorticosterone was measured in the urine (Mason *et al.*, 1968a; Sassenrath, 1970). Cortisol can be measured in urine and compared with levels of creatinine which is excreted at a relatively constant rate (Novak and Drewsan, 1989). Urine samples can sometimes be collected with minimal disturbance to the individual, but the collection must occur some considerable time after the event whose effects are being studied.

A major problem with all attempts to measure adrenal cortex responses in plasma is that the action of taking a sample often evokes a considerable response. However, there is a much greater delay before glucocorticoids are released into the bloodstream than occurs before adrenaline or noradrenaline are released from the adrenal medulla. In most species the delay before glucocorticoids are released is at least 2 minutes, so the effects of a particular treatment can be measured if a blood sample is collected within 2 minutes of the beginning of the blood sampling procedure. A further problem is that there is a cycle in baseline adrenal cortex activity, and the magnitude of the response when an animal is stimulated can vary according to the stage of this cycle. For example Seggie and Brown (1975) found that the corticosterone response to handling was two and a half times greater in rats at the peak of the baseline cycle than in those at the trough.

In some situations a saliva sample can be taken and assayed using ELISA (Cooper *et al.*, 1989). The quantity of cortisol in saliva is less than that in plasma so a more sensitive assay is needed. Plasma glucocorticoids exist in free and protein-bound forms. Only the free form is present in saliva but this is probably the most relevant when assessing responses to environmental difficulties. Salivary cortisol levels increase in response to ACTH injection in sheep (Fell, Shutt and Bentley, 1985) and in pigs (Mendl *et al.*, 1991, 1992), and are not affected by salivary flow rate (Fell and Shutt, 1986). Levels of salivary cortisol have been found to increase in similar circumstances to those in which plasma cortisol increases in humans (Riad-Fahmy *et al.*, 1982; Vining *et al.*, 1983a,b), sheep (Fell, Shutt and Bentley, 1985; Fell and Shutt, 1986), pigs (Parrott and Misson, 1989; Parrott, Misson and Baldwin, 1989) and dogs (Vincent, personal communication). However the data of Trunkfield (1990) on calves indicated only a general relationship and much variation.

The work of J.W. Mason on the effects of a variety of situations on urinary glucocorticoid levels in monkeys has already been discussed (Section 4.2.2). More recent work has involved assessing the effects of various treatments on the production of species-specific glucocorticoids. In primates, dogs, cats and most ungulates the predominant glucocorticoid produced is cortisol; in rodents and chickens, it is corticosterone. However, in some species, for example pigs, both are produced (Sharman pers. comm., Broom and Guise in prep) and this has often been overlooked. The magnitude of the adrenocortical response also varies from one breed to another in domestic animals, for example, the corticosterone response in poultry to handling (Freeman and Flack, 1980). This fact and the evidence for wide individual variation in adrenal, as in other coping responses, have highlighted the need for careful control of genetic diversity in subjects and for the use of large sample sizes in studies of the effects of different treatments on adrenal cortex responses.

Situations which we would expect to be painful are sometimes, but not always, associated with increased plasma glucocorticoid levels. Silver (1982) found such increases after surgical treatment for tendon damage in horses, and several studies have reported increases in cortisol levels associated with behavioural responses to operations such as tail-docking, castration, and mulesing (Section 5.1.3) in sheep (Shutt *et al.*, 1987; Mellor and Murray, 1989; Kent, Molony and Robertson, 1991; Wood and Molony, 1991). In most of these studies, the severity of the procedure was correlated with the magnitude of the adrenal cortex response (Table 5.2).

Table 5.2 Effects of surgical procedures on lambs (from Shutt *et al.*, 1987)

	Cortisol (n mol l^{-1})
Control	87
Tail dock	136
Castration	171
Mulesing and tail dock	187*
Mulesing, tail dock and castration	232*

*Cortisol levels still high 24 h after mulesing

Further evidence for links between adrenal cortex response and pain comes from the findings that analgesia can substantially reduce the response, for example in new-born human infants undergoing surgery (Anand, Sippell and Aynsley-Green, 1987; Anand *et al.*, 1988). Such results have led to a general move to use analgesics after minor operations on new-born babies in some countries. Links between predictably painful situations, adrenal cortex response and suppression of this response by a proven analgesic are particularly useful in the assessment of pain, but Mellor (personal communication) has reported adrenal cortex responses associated with tissue damage even when an anaesthetic was used. It seems that the damage itself, even if not perceived, can cause an adrenal response.

Various forms of laboratory animal handling, for management or experimental purposes, elicit adrenal cortex responses. Kvetnansky *et al.* (1978) found that handling for 30 seconds doubled plasma glucocorticoid levels in rats and significant increases also occurred in rats when their cages were moved from a rack to the floor or a table (Gärtner *et al.*, 1980). Mice and rats show elevated plasma corticosterone levels when put into a novel environment and the extent of elevation is related to the degree of novelty, as shown in Figure 5.1 (Hennessy and Levine, 1978; Hennessy *et al.*, 1979; Armario *et al.*, 1986).

Handling and other treatment causes increases in plasma cortisol levels in trout (Pickering and Pottinger, 1985). Handling and transport of farm animals also results in an adrenal cortex response. Broom, Knight and Stansfield (1986) reported that plasma corticosterone levels in hens from a battery house were 4.3 ng ml^{-1} 5 minutes after the normal, rough handling which occurs before transport to slaughter, but only 1.45 ng ml^{-1} after gentle handling. Freeman *et al.* (1984) found that untransported broiler chickens had a plasma corticosterone level of 1.3 ng ml^{-1}, but the level after 2 hours of transport was 4.5 ng ml^{-1} and after 4 hours of

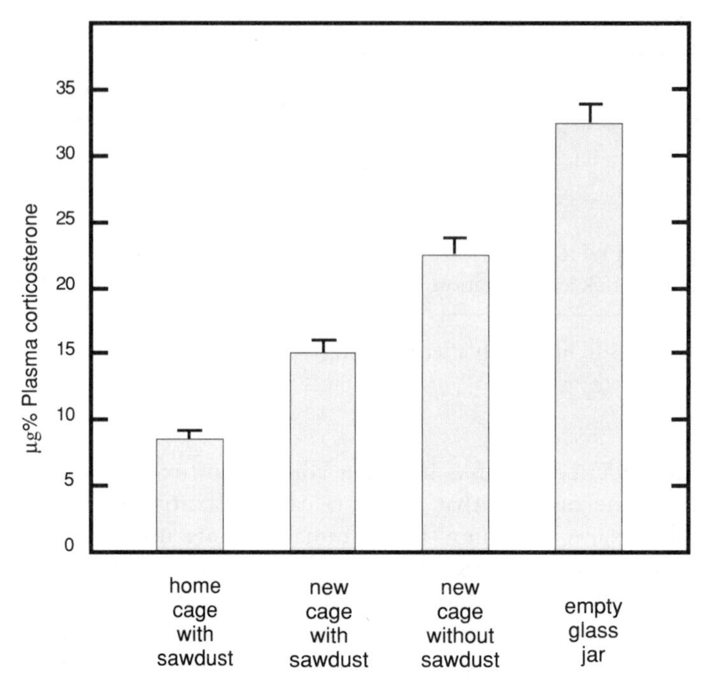

Figure 5.1 Plasma corticosterone in mice in their home cage, and shortly after exposure to novel situations with different degrees of difference from the home cage (modified after Hennessy and Levine, 1978).

transport was 5.5 ng ml^{-1}. Kent and Ewbank (1983, 1986) studied the effects of transport on calves, and took account of the fact that long journeys include effects of food deprivation as well as other physical and social conditions. They compared calves on an 18-hour journey with calves simply deprived of food for the same period. The peak value for plasma cortisol occurred 10 minutes after the start of the journey (Fig. 5.2), but it was also significant that the cortisol level of the control calves remained low during the 18 hours but increased when food was provided. The increase may have been associated with increased metabolism following eating or may have been a consequence of the circumstances in which the food was presented.

Laboratory experiments can lead to substantial adrenocortical responses. The plasma glucocorticoid levels of rats which were restrained in a plastic tube were much higher than those of animals put in a novel environment (Armario *et al.*, 1986). Restraint of monkeys in a chair also caused plasma cortisol increases (Goncharov *et al.*, 1979; Hayashi and Moberg, 1987). Electric shock applied to the feet elicits a substantial adrenal cortex response in rats even if current levels are thought to be low (Natelson *et*

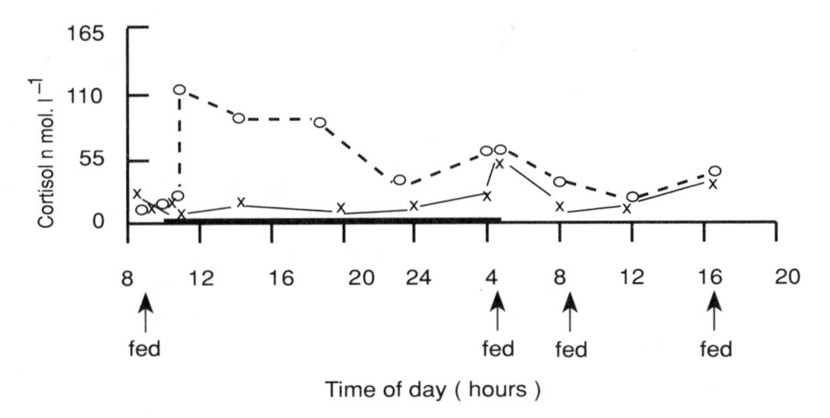

Figure 5.2 Changes in plasma cortisol during road transportation of three-month-old calves. Values for transported calves (o) were significantly different from values for calves starved for the same period (x) (after Kent and Ewbank, 1986).

al., 1981, 1987; Maier *et al.*, 1986). After once being in the box in which the shocks were administered, substantial responses were produced as soon as the rats were again put into the box (Kant *et al.*, 1983). Even exposure to the odour of rats which had experienced shocks resulted in adrenal cortex responses (MacKay-Sim and Laing, 1980). Infant monkeys separated for two-hour periods from their mothers, mice attacked after being put in a new colony, and rats or guinea pigs involved in fights all showed elevated plasma glucocorticoids (Hennessy, 1986; Henry and Stephens, 1977; Koolhaas *et al.*, 1983; Sachser, 1987).

In general, it is clear that measures of glucocorticoid responses are of considerable value when assessing short-term effects on animals. The various problems and variables mentioned in this section must be taken into account during interpretation, however.

5.2.4 Other hormones

Various reproductive hormones are affected by exposure to difficult conditions. We do not know in detail how the changes in these hormone levels help animals to cope, or how they may indicate likely failure to cope. We consider them as welfare indicators only, because the changes in hormone levels are associated with situations which appear to pose problems for the individual, and in which there are other indicators of poor welfare.

One hormone much affected is prolactin, which can be measured in plasma using radioimmunoassay. A rise in levels occurs within 1 minute of stimulation and there is a peak 8–15 minutes later in rats (Seggie and Brown, 1975; Gärtner *et al.*, 1980). The delay before return to basal levels is 20–60 minutes depending upon the extent of the effect on the individual. The magnitude of the prolactin response depends upon the basal levels, which depend in turn on the sex of the individual and, in females, on the stage in the oestrus cycle. During pro-oestrus basal prolactin levels are high, but during dioestrous they are low, so an increase in response to difficult conditions is possible (Riegle and Meites, 1976; Turpen, Johnson and Dunn, 1976). Prolactin responses in rats have also been reported to decline with increasing age (Milenkovic *et al.*, 1984). Quite small disturbances, for example 5 seconds of handling, blood sampling, intraperitoneal saline injections or cage movement can elicit plasma prolactin increases, so it can be a sensitive measure of welfare (Krulich *et al.*, 1974; Seggie and Brown, 1975; Gärtner *et al.*, 1980). For restraint and foot-shock, prolactin levels are proportional to the magnitude of the treatment (Kant *et al.*, 1983a; Kant, Mougey and Meyerhoff, 1986). In human and other primates, prolactin increases do not follow exposure to difficult conditions (Eberhart, Keverne and Meller, 1983; Brooks *et al.*, 1986). Parrott (1990), working with sheep, found that restraint caused elevation of plasma prolactin, but isolation, which results in cortisol increase, had no effect on prolactin levels (see also Parrott *et al.*, 1987).

Levels of luteinizing hormone (LH) in plasma are also elevated after a variety of treatments but, like prolactin, the response depends on the stage of the oestrus cycle (Turpen, Johnson and Dunn, 1976). Handling, a novel environment (Euker, Meites and Riegle, 1975), restraint (Krulich *et al.*, 1974) and noise (Armario *et al.*, 1984a) all lead to increases, but the LH response appears to be less sensitive than the prolactin response. Restraint of monkeys can also lead to an LH increase in plasma (Hayashi and Moberg, 1987). Follicle stimulating hormone and testosterone are of little use as welfare indicators because they may increase, not change or decrease in situations which are otherwise recognizable as difficult for animals.

The glucocorticoids are obviously the hormones most extensively involved in metabolic adjustments to difficult conditions, but other hormones also may reflect changes in the environment. Glucagon is involved in the regulation of plasma glucose levels but has been measured in only a few studies of responses to trying conditions. Freeman and Manning (1976) found increased levels in poultry following handling, intramuscular injections or cardiac puncture. Transient increases followed tendon surgery in horses (Silver, 1982) and increases related to the extent of surgical trauma have been reported in humans (Anand, Sippel and Aynsley-Green, 1987; Anand and Aynsley-Green, 1988). Thyroid stimulating hormone appears to be of little value as an indicator of stressful conditions. Growth hormone is sensitive to environmental conditions but is so variable during the day that it is of little use.

Atrial natriuretic peptide (ANP) is produced by muscle cells in the cardiac atria in response to atrial stretching and under certain conditions to hormones from the adrenal cortex and adrenal medulla (de Bold, 1985; Brenner *et al.*, 1990). Levels of ANP in rat plasma increased according to the degree of foot-shock administered to the animal. This could be a useful welfare indicator.

The naturally occurring opioids, including ß-endorphin, are produced at the same time as ACTH in some circumstances. For this reason alone, their levels in plasma are of interest during welfare assessment, even if it is still uncertain whether these measures reflect any analgesic function or indicate the animal's success in coping. Restraint of monkeys and rats leads to increases in plasma ß-endorphin (Mueller, 1981; de Souza and van Loon, 1985; Kalin *et al.*, 1985a). Foot-shock also causes elevation in plasma ß-endorphin (Rossier *et al.*, 1977; Millan *et al.*, 1981; Kant *et al.*, 1983b). Extreme pain may be suppressed because of the secretion of naturally occurring analgesic opioid peptides such as ß-endorphin and the enkephalins (Hughes *et al.*, 1975) but it is not easy to take account of this fact when making measurements of welfare following short-term problems. Studies of the densities of mu, delta and kappa receptors in various parts of the brain may give information about welfare after more long-lasting problems (Zanella, Broom and Hunter, 1991, 1992) (Section 6.8).

.2.5 Neurotransmitters

Responses to novel, threatening and alarming environmental events involve dopaminergic and noradrenergic systems in the brain (Smelik, 1987). After being involved in such activities, dopamine and noradrenaline may be depleted in parts of the brain, or there may be increases in the levels of metabolites of these neurotransmitters. Over a longer period, increased levels resulting from greater than normal synthesis may be evident. These changes are most readily detected after death, although experimental situations in which push–pull cannulae sample minute quantities of fluid from different brain regions can be used (Wuttke *et al.*, 1984). Exposure of laboratory rats to severe treatment, such as inescapable electric shock, rapid rotation or swimming in water at 4 °C for 3 minutes, can result in depletion of hypothalamic noradrenaline (Swenson and Vogel, 1983; Roth *et al.*, 1982). The handling and transport experienced by domestic hens when removed from their cages and taken to a slaughterhouse also results in some hypothalamic noradrenaline depletion (Broom, Knight and Stansfield, 1986). The major noradrenaline metabolite, 3-methoxy-4-hydroxyphenyl-ethylene-glycol (MHPG), appears in urine in man at levels which are proportional to the difficulty of conditions experienced (Rubin *et al.*, 1970). However, in rats this relation is not so simple, since MHPG is produced peripherally as well as centrally (de Met and Halaris, 1979) and levels are affected by diet (Sedlock *et al.*, 1985).

The levels of dopamine in the brain of animals which have experienced difficult conditions vary according to the nature of the stimulus, so the value of such measurements is dubious. Anisman, Pizzino and Sklar (1980) reported depletion of dopamine in the arcuate nucleus of the rat hypothalamus after various unpleasant and inescapable stimuli. The dopamine metabolites 3,4-dihydroxy phenylacetic acid (DOPAC) and homovanillic acid (HVA) increased in the brainstem and hypothalamus of rats subjected to swimming in cold water or rotation (Roth, Mefford and Barchas, 1982) but the level of dopamine itself remained constant.

Some changes in levels of other neurotransmitters and their metabolites have been reported after short-term problems were encountered. In general, however, such measurements can only be made in situations in which the animal will be killed. In other circumstances, welfare assessment could be helped by measuring levels of a metabolite such as MHPG in plasma or urine, provided that all of the sources of this substance have been investigated.

5.2.6 Enzymes and metabolic products

Measurements of most plasma enzyme activity are more appropriate for long-term than for short-term welfare assessment because it takes time for additional enzyme to be produced following an environmental change. However, some enzyme levels change rapidly enough to provide a useful indicator of short-term problems. Renin is produced in the kidney and is involved in the regulation of water balance and blood pressure. Its levels in the plasma are interrelated with sympathetic nervous system activity. In dogs, adrenaline injection increased plasma renin activity and the ß-blocker propranolol, which partly inhibits sympathetic activity, and reduced plasma renin activity (Johnson, Shier and Barger, 1979). Surgery increased plasma renin activity in man (Anand, 1986), as did immobilization, foot-shock, or immersion in cold water in rats (Paris *et al.*, 1987). Maximum levels were found 12 minutes after the disturbing stimulus and the levels declined 10–20 minutes after the end of the experience.

Various proteins are found in plasma at much higher levels after tissue damage and some are released into the bloodstream at the time of unpleasant experiences which do not cause tissue damage (Adams and Rinnie, 1982). Transient increases in the plasma levels of the liver enzymes alanine aminotransferase and aspartate aminotransferase were found in baboons after capture (Steyn, 1975). However, levels of these enzymes and of alkaline phosphatase were not found to change in rats following exposure to a novel situation which caused a plasma cortisol increase (Gärtner *et al.*, 1980). Creatine kinase, produced from heart and skeletal muscle after damage or vigorous exercise, increased in rabbits following blood loss (Bacou and Bressot, 1976) and in rats kept in metabolism cages with a wide-mesh floor (Frölich *et al.*, 1981). It was suggested that the high values in the rats resulted from the muscular effort associated with trying to balance on the floor wire, as plasma corticosterone levels did not increase. Calves kept in outdoor yards where they could exercise had higher plasma creatine kinase levels than those kept in very small pens (Friend, Dellmeier and Gbour, 1985). It remains to be demonstrated whether levels of this enzyme can be used as a welfare indicator, but a recent study (Trunkfield *et al.* in preparation), demonstrating that creatine kinase levels in calves were higher after injection with an antibiotic which causes some tissue necrosis, suggest that the measure is useful in combination with other measures.

Another enzyme which leaks from muscle into the bloodstream in certain situations is a form of lactate dehydrogenase (LDH). There are five isoenzymes of LDH; LDH5, which occurs mainly in striated muscle and liver, increases in plasma after muscle damage or exposure to disturbing conditions in pigs. A high incidence of pale, soft, exudative meat (PSE) in slaughter pigs is associated with high levels of LDH5. Increases in levels of plasma LDH5 in cattle after transport have been reported by Mormède *et al.* (1982) and Trunkfield (1990). Increases in total LDH levels after capture, handling or transport have also been reported in baboons (Steyn, 1975), cattle and pigs (Moss and McMurray, 1979), and these are probably due mainly to increases in the LDH5 form. Plasma LDH5 also increased in park deer after capture (Jones and Price, 1990). Of particular interest in this study was the fact that increases continued in animals lying quietly with their heads covered. Hence it seems that the release of this isoenzyme into plasma cannot be just a consequence of exercise, but is a response of the animal to a disturbing situation. The increase in the enzyme takes many minutes to occur and levels remain high for some hours, so it could be a more useful indicator of welfare than the more transient changes in hormones.

Plasma glucose levels are increased following the secretion of adrenal medullary and cortical hormones, but they can also be reduced by vigorous activity. Plasma glucose levels increased in rats following: electric shock (Natelson *et al.*, 1977); movement of their cage from the rack to the floor, or exposure to ether (Gärtner *et al.*, 1980); and transfer into a new cage (de Boer *et al.*, 1989). However, Quirce and Maickel (1981) found that although plasma glucose levels of mice were increased initially by either immobilization or low ambient temperatures, this increase was followed by a decrease below the baseline. A transient increase in plasma glucose for 10 minutes, followed by a decrease over several hours, was also found in calves during transport (Kent and Ewbank, 1986). Overall it seems that measurements of plasma glucose are of little value as welfare indicators unless repeated samples of plasma are available, and the obtaining of these may cause further changes in adrenal activity and glucose production.

Other metabolites such as free fatty acids and cholesterol have been measured following exposure to a variety of unpleasant situations, but the responses do not appear to be consistent and are therefore of little value as indicators of welfare.

5.2.7 Muscle and other carcass characteristics

When farm animals are handled or transported, biochemical changes, especially those associated with glycogen metabolism, occur in muscle (Hails, 1978; Tarrant, 1981; von Mickwitz, 1982). When pigs are subjected to such disturbance before slaughter, there can be very rapid glycolysis, with consequent high production of lactic acid and fall in pH (Lendfers, 1970; Augustini, Fischer and Schön, 1977). As a result, the water-binding capacity of muscle protein declines, water leaks out of the meat and the colour becomes paler and greyer. The incidence of the pale, soft, exudative (PSE) meat produced gives information about the welfare of the animals in the period shortly before slaughter, but the substantial genetic variation in the likelihood of PSE meat production must be taken into account. Such meat is of considerably lower palatability and, hence, value. Dark, firm, dry (DFD) meat is produced if glycogen reserves are depleted before death so that little lactic acid can be produced in the muscles after death and the pH remains high. It is less attractive to shoppers choosing meat and, although it is usually more tender than normal meat, its value is lower. In cattle, fighting, disturbances caused by mixing animals from different groups, and adverse weather can lead to DFD meat after slaughter (Tarrant, 1981) and the same can be true in pigs. The presence of DFD meat also provides evidence about the welfare of the animals prior to slaughter.

Injuries to animals following handling or transport are important indicators of their welfare. Guise and Penny (1989a,b) reported that skin blemishes on carcasses of pigs after slaughter, including scratches and bruises, could be related to the conditions experienced by the animals during their last few hours. For example, the number of blemishes was increased if animals were mixed with strangers. Bone breakage can also occur, especially in poultry. Gregory and Wilkins (1989) dissected more than 3000 hens from battery cages in the UK and found that a mean of 29% had broken bones before they reached the water-bath stunner in the slaughter line. Bone breakages and other injuries are important indicators of poor welfare, the effects ranging from very slight to extremely severe. They can be used in combination with indicators of pain and discomfort.

5.3 USING INDICATORS TO EVALUATE WELFARE

When measures are made of the attempts of animals to cope with short-term problems, figures obtained for changed behavioural response, increased heart rate, or the extent of any increase in a hormone or enzyme, have to be interpreted. What do these figures tell us about welfare? Can the various measures be interrelated? Is a certain increase in heart rate equivalent to a certain degree of inhibition of normal behaviour? Some responses occur only in more extreme situations, whereas others appear when only a slight disturbance occurs. Nonetheless, levels of one measure may still be equated with levels of another. On the other hand, one kind of coping response may be an alternative to another. Our knowledge of these matters is lamentably poor. We may also be failing to use some potential welfare indicators because we do not know they exist. The questions posed above will now be considered in turn.

Some behavioural responses to difficult situations are not related in intensity to the degree of difficulty encountered, but are all-or-nothing. Alarm calls or displays are usually either shown or not shown. However many behavioural responses are quantifiable. We can measure the number of distress calls, the frequency of kicking at a localized pain source, or the duration of a response. In general it seems logical to assume that a more intense or prolonged response means a greater problem for the individual concerned. Hence in Baldock and Sibly's (1990) study of heart rate in sheep (Section 5.2.1), animals were presumed to be more disturbed, as indicated by higher heart rates, when confronted by the strange man with a dog than by the man alone or by transport. There must be a maximal level of heart rate, however, so it is not possible to differentiate among responses to stimuli which, although differing in their overall effect on the animal, all elicit the maximal response. In this circumstance, an additional measure is used, such as the delay before the high heart rate returns to resting levels (Duncan, 1986). This measure is particularly useful because of the individual variation in heart rate response; with this technique, as in Baldock and Sibly's study, animals can be used as their own controls.

Measures of adrenal cortex response also show a graded response over a range of increasing difficulties and have a maximal response (Stephens, 1980), so the same arguments apply. For some physiological measures, however, the delay before return to resting levels may be either a function of the maximum level reached, in which case no extra information is obtained, or may be unrelated to the level reached or to the effects of conditions on the individual. In order to use any one of the measures discussed in this chapter as an indicator of welfare, the significance of different magnitudes of the effect and of different patterns of return to normal must be studied in the species concerned.

Any simple measure, such as the occurrence of escape behaviour, gives information about how poor the welfare of an individual is. But a lower level of that response in another individual does not necessarily mean that the welfare of the second animal is better. The animal could be using a different coping method, or could be affected in a different way from the first. Duncan and Filshie's (1979) finding that a supposedly docile strain of hens, which showed a small behavioural response to human approach, showed a much greater heart rate response than a supposedly 'flighty' strain (Section 5.2.1) is a good example of this. The examples given in this chapter of individuals showing either a more active or a more passive coping response when confined with an aggressive individual (Section 5.1) also show that care must be taken in assessing welfare. A passive withdrawal response may indicate very poor welfare, as may an active avoidance response, so a simple measure of activity levels would give quite erroneous information about the magnitude of the effects on the various individuals. As discussed in Section 5.1, different sets of behavioural and physiological responses can be shown in identical situations by different individuals. The general problem of how to assess welfare when individuals vary in their responses to difficult conditions has been discussed by Broom (1986b, 1988b). A single measure can indicate that welfare is poor but a combination of measures is preferable if valid comparisons of conditions affecting the welfare of animals are to be made.

Problems therefore still arise when it comes to deciding which measures are most important. Is a large adrenocortical response worse than a large heart rate response, a large behavioural response, or a certain kind of injury? Some indication of the importance of the different measures is included in this chapter, but better categorization and evaluation of measures is needed (see Chapter 8).

5.4 SHORT-TERM WELFARE PROBLEMS AND CONCEPTS OF STRESS

As explained in Section 4.1.4, the term stress is used when we can see that the environmental effect on an individual is reducing, or is likely to reduce, the fitness of the individual. Throughout this chapter the word welfare has been used because, although it is often clear that the individual is having difficulty in coping, it is not clear that there will be an effect on fitness. A brief period of elevated heart rate, adrenal cortex activity, or emergency behaviour indicates that welfare is poorer in a particular animal than in an individual which does not have such responses, but they may have no effect at all on life expectancy, the period before the next breeding or the number of young produced at next breeding. However, an extreme response or a series of relatively extreme responses may affect fitness. In an extreme situation, a substantial startle response could lead to a heart attack, or to some cardiac tissue damage which makes subsequent heart attacks more likely to be fatal.

As discussed further in the next chapter, high levels of adrenal activity, even for short periods, can cause sufficient immunosuppression for a pathogen attack to be successful, whereas such an attack would otherwise be warded off by the immune system defences. Likewise a behavioural response to some short-term problem could make an individual more vulnerable to predator or parasite attack, and thereby reduce its fitness.

Hence some responses to short-term problems will reduce fitness, even if the majority of such responses do not. The word stress can be used, somewhat loosely, in relation to many circumstances which result in poor welfare, but there are many circumstances in which welfare is affected but there is no stress. Sometimes a disturbing situation which is, in itself, very unlikely to reduce fitness can be very difficult for the individual to cope with, and so there is a considerable effect on welfare. Human phobias can be examples of this. The fear of spiders, in countries where there are no dangerous spiders, may mean that a brief encounter with a spider is disturbing to the person for a few seconds or minutes, so welfare is temporarily poorer, but no effect on fitness occurs unless the response itself adversely affects fitness.

It is desirable that, for human and other species, the relationships between short-term problems and effects on fitness be more thoroughly understood. Firstly, there is the question raised above of the links between extreme, but brief, periods of poor welfare and ultimate effects on fitness. Secondly, we need to consider welfare in relation to time, and the relative importance of short- and long-term problems. It could be assumed that the overall effect of a condition or treatment is best assessed by measuring how poor welfare is (PW), how long this continues (t), and then quantifying the effect as (PW×t). In practice, welfare would certainly vary over this time, so the overall effect of plotting the extent of poor welfare against time would be represented by the area of divergence from the neutral line (Fig. 5.3). Such a calculation could be used for comparing two short-term problems (Figs. 5.3a and b), or a short-term and a long-term problem (Figs. 5.3a or b with c).

Such a calculation is worthwhile in comparative studies but we cannot be sure that the simple multiplicative relationship shown is correct. There will also be difficulties when different measures must be used at different times. We must try to quantify the effects of conditions and treatments on welfare, and calculations like those shown would be a useful start in this attempt.

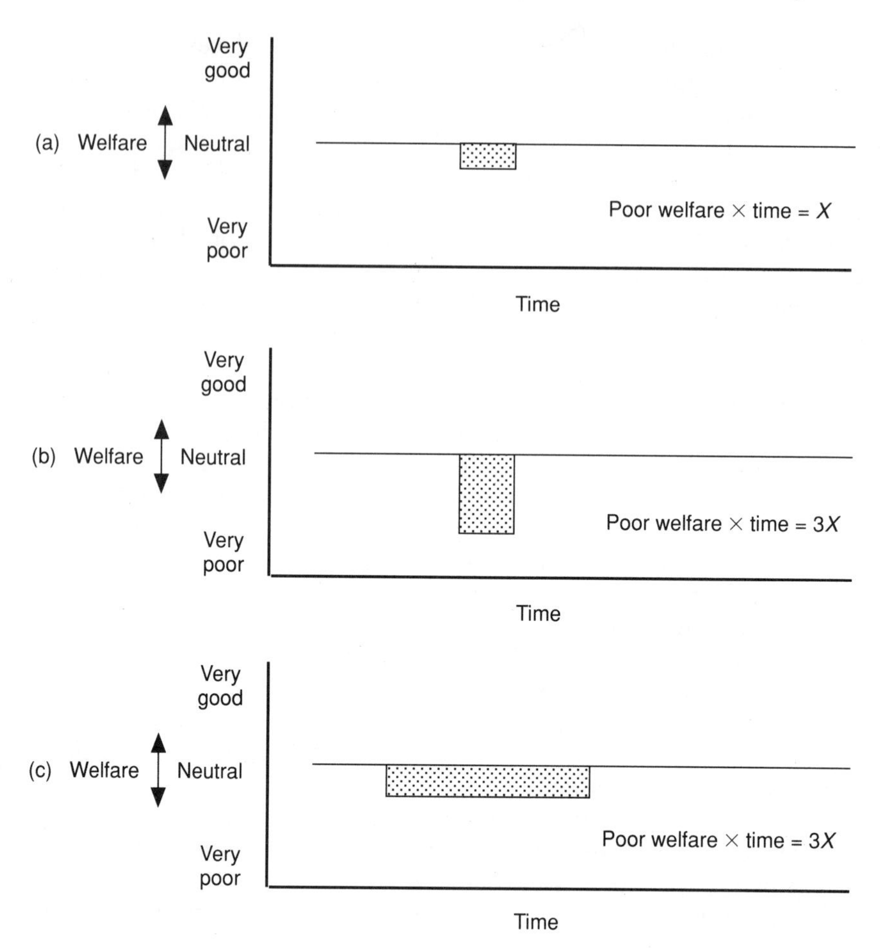

Figure 5.3 Relationships between poor welfare and time. The effects of two levels of poor welfare (PW) and two durations of environmental conditions on an individual are shown.

Chapter 6
Assessing welfare: long-term responses

Most of us would accept that weeks or months of hell are worse than minutes or hours. In biological considerations of stress this difference is important because animals use different coping methods when problem situations are prolonged rather than brief. Persistent problems are also often the major causes for concern when we are judging the welfare of domestic animals, or people.

Assessing poor welfare when problems are prolonged involves measuring a range of variables, including fitness in the biological sense, which indicates stress, and the difficulty an individual has in coping with its environment. These measures overlap with some of those for short-term problems described in Chapter 5, but because individuals adapt relatively rapidly to most short-term problems, different sets of measures are required for assessing short- and long-term problems.

Our knowledge of the responses of animals to long-term welfare problems is limited. Recent research, such as that on changes in immunosuppression and plasma concentrations of naturally occurring opioids, may uncover new coping methods or consequences of failure to cope. In the meantime, we must use the measures we have, and these are considered in this chapter. The first variables to be examined are those that reduce individual fitness, that is, those affecting reproduction and survival, and which include some of the signs of poor welfare that are most obvious to the lay person. These provide definitive evidence that the individual is stressed.

6.1 REDUCED REPRODUCTIVE SUCCESS

If an individual is so deprived of food or disturbed by its environment that it is unable to reproduce when given the opportunity, its welfare is poorer than that of another individual which can reproduce. In wild animals, and also in neglected domestic animals or people, starvation can delay or prevent reproduction. Even when the food supply is adequate, an animal may be so unsettled by its living conditions that it does not reproduce when given the opportunity. Many species of animals are not able to breed in the poor conditions in some zoos, and some species rarely breed in any zoo.

Relative reproductive performance can be used to compare the welfare of animals in different conditions though, in practice, conditions have to be very poor before reproduction is affected in most animals, including humans. Body resources are often apportioned to reproductive effort even at the expense of basic body maintenance. Farm animals, having been selected for many generations for good reproductive performance, commonly maintain normal reproductive performance even under relatively difficult conditions.

In farm animals, a failure to reproduce or a delay in reproduction is usually either a consequence of an inadequate diet or is associated with some behavioural abnormality (Lindsay, 1985). Cows, mares, ewes and sows may be disturbed by the proximity of dominant individuals, or may be attacked by them, and so fail to show normal behavioural oestrus (Fraser and Broom, 1990). This results in failure to breed and hence reduces fitness. Although most frequently seen in subordinate individuals or animals subjected to frequent changes in social group, lack of behavioural oestrus may also be caused by excessive noise, temperatures above 30 °C and extreme weather conditions (Hurnik, 1987).

Male farm animals may fail to show appropriate sexual behaviour, and indeed may be unable to show it, in circumstances which create apprehension (Fraser, 1960; Fraser and Broom, 1990). Some males show an interest in a potential sexual partner, but are disoriented during copulation or fail to achieve intromission if they have been deprived of suitable social contact during their early lives (Hemsworth *et al.*, 1978, Zenchak and Anderson, 1980; Beilharz, 1985; Price, 1985a). Females can also have poor reproductive skills owing to disturbances or inadequacies in their early experience, as reported in monkeys (Harlow and Harlow, 1965), cows (Broom, 1982), mares (Houpt, 1984), and ewes (Arnold, 1985). Disturbed mothers of some species, including pigs, may eat their young after giving birth (Sambraus, 1976; Fraser and Broom, 1990).

6.2 LIFE EXPECTANCY

If the life expectancy of animals in one set of conditions is two years whilst that in another is six years, we could conclude that the welfare is poorer in the first case than in the second. This general idea has been presented by Hurnik and Lehman (1988) and by Broom (1991a, in press). However, some reservations should be made about the comparison. Individuals which live for a shorter time may die early because they are diseased, and these animals are unquestionably stressed. In other cases the metabolic rate or pace of life in the shorter-lived animals may be greater. Some people argue that to live at a faster rate for a shorter time is no worse than to live longer at a slower rate. But this is not the view of the majority of people, who regard those domestic animals living at a higher rate for a shorter time as being the more stressed, particularly when conditions are imposed rather than being chosen by the animals.

Reduced life expectancy due to sub-optimal living conditions can occur in the wild, for example in trout (Pickering, 1989a). It is also known to occur when inappropriate animals are kept as pets or in zoos. For example, several small species of tortoise which can live for longer than 20 years in the wild have an average life expectancy of only 2 years in captivity (Warwick, 1989). Wild-caught birds often have short life expectancies in captivity, and primates kept as pets seldom live as long as those kept in good zoo environments or those living in stable groups in the wild. Perhaps the best known examples of animals which do not live long in zoo conditions are cetaceans: whales and dolphins kept in pools no larger than human swimming pools live much shorter lives than those kept in larger marine enclosures or living in the wild.

Meaningful comparisons of life expectancy of husbanded and wild animals are difficult because wild animals have to contend with predators, parasites and pathogen challenges which are avoidable in captivity. An estimate of life expectancy in the wild should be made by considering individuals which are not eaten by predators or severely affected by diseases and parasites. The question of whether animals have better welfare while living a longer, safe life in protected custody or a shorter, more hazardous life in the wild is a philosophical matter which will be referred to again in Chapter 8.

The welfare of farm animals is effectively measured by 'potential' rather than 'real' life expectancy, since these animals are usually killed for human consumption before they die naturally. Dairy cows are kept by most farmers until they cease to give a high milk yield, after which they are culled. Others are culled because they fail to become pregnant, or because they are lame or have some other disability. Some of these factors are similar to those which would result in premature death in the wild, whilst others, such as poor ability to conceive and rear young, impair biological fitness. Early culling or death for such reasons on a particular farm indicates conditions of poorer welfare than on another farm managed so that the animals live longer with less necessity for culling.

Many farmers have reported that the average age to which their dairy cows lived during the 1980s and 1990s was less than that of cows in earlier years. They link this with the high production rates of modern dairy cows. If a cow exhibits high feed-conversion efficiency and a high metabolic rate, as a consequence of a high protein diet and selection for such efficiency over many generations, longevity may be reduced. Evidence for such a change in modern dairy cows is indicated by the number of cows in Denmark being sent to rendering plants, which normally occurs after they have died on the farm. Using such data, Agger (1983) reported that the life expectancy of dairy cows was halved between 1960 and 1982 (Fig. 6.1).

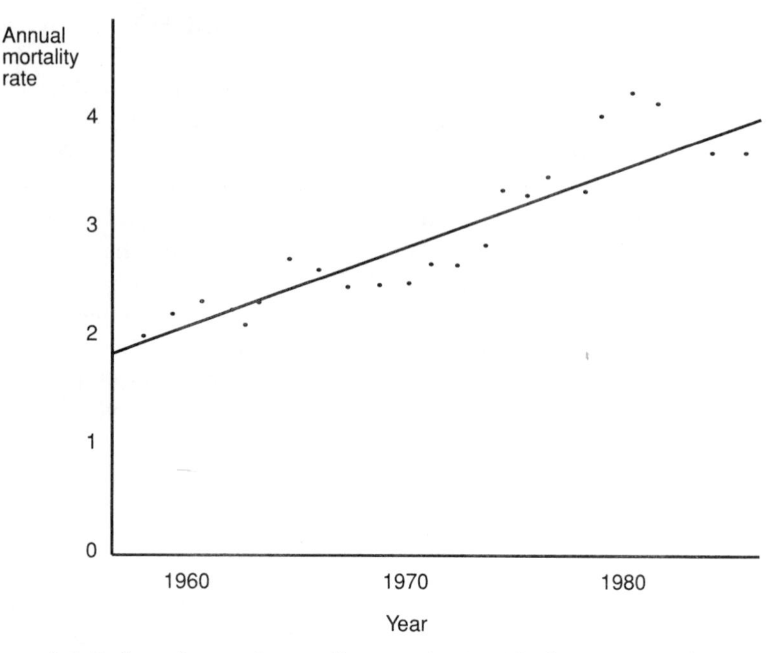

Figure 6.1 Index of annual mortality rate in Danish dairy cows (after Agger, 1983).

This trend may well continue as milk production rates go on increasing. Wider use of bovine somatotrophin to increase milk production would probably accelerate the fall in life expectancy.

6.3 WEIGHT CHANGES

An early sign that an individual might not reproduce, or is ill and likely to live for a shorter time, is an interruption to its growth or, in grown animals, a loss of weight. In both wild animals brought into captivity and in domestic animals, weight loss in adults or lack of weight gain in juveniles usually indicates severe conditions for the animal.

The tree shrews studied by von Holst (1986) lost weight and died if they were confined with an individual which had beaten them in a fight. Young rats which consistently won fights gained weight, but those which were losers lost weight (van der Poll *et al.*, 1982). In both cases the animals continued to feed, so presumably differences in metabolic rate associated with increased adrenal activity caused the weight loss.

Many unpleasant conditions, if persistent, can affect body weight. Daily exposure to cold, forced exercise, foot-shock and immobilization have led to reduced body weight in rats (Taché *et al.*, 1978; Konarska *et al.*, 1989a,b). In the experiment of Taché *et al.*, immobilization had the greatest effect, and weight loss was evident after three days. Rats subjected to various laboratory procedures, such as movement to a new room, change from a pelleted to a powdered diet, change from groups of three to single housing, oral dosing with water and daily cage change over a period of 44 days showed a net loss in weight, whilst undisturbed controls gained 9% in body weight over the same period (Steinberg and Watson, 1960). Housing in a more complex environment resulted in a greater weight gain by mice, even though these animals were more active than mice kept in bare cages (Chamove, 1989a,b). The complexity of the environment was probably also responsible for the greater weight gain in Syrian hamsters kept in groups, rather than singly, by Arnold and Estep (1990).

Stability and size of groups are apparently important. Edens (1987) studying quail, and Mormède *et al.* (1990) studying rats, both found that frequent movement to a new social group reduced weight gain. Very crowded conditions reduced weight gain in rats studied by Armario and Jolin (1989).

Weight gain is monitored in many young farm animals. Efforts are made to minimize any reduction in growth rate, especially at weaning when there are changes in location and social companions, as well as in diet. Piglets are commonly separated from their mothers at 3–5 weeks, well before natural weaning at 9–16 weeks. When piglets are put into a new pen and mixed with strangers there is an increase in fighting and a drop in food intake, so weight may drop or fail to increase at the previous rate.

6.4 CARDIOVASCULAR AND BLOOD PARAMETERS

Heart rate is a useful measure of welfare in the short term, but of little value when comparing long-term conditions, such as the quality of housing. However, long-term conditions can affect changes in heart rate or blood pressure which occur in test situations.

Blood pressure measurements must be interpreted carefully, since taking the measurement can have an effect on the animal. O'Neill and Kaufman (1990) reported that intra-arterial catheterization of rats resulted in decreased food intake and body weight loss. Baboons can be trained to hold out forelimbs for blood pressure measurement using an oscillometric monitor (Turkhan *et al.*, 1989), and such a technique may be possible with other animals habituated to human contact and hence less disturbed by it. Some implanted transducers are not major long-term impositions on the animal but do require surgery for implantation. Cuffs on the forearm of a primate, the ear of a rabbit or the tail of a rodent are non-invasive and seem to be practicable methods of blood pressure measurement, provided that the individual is not disturbed by the restriction of movement.

Frequent confrontations and fights between mice over a period of months resulted in a permanent increase in basal blood pressure from 125 to 150–175 mmHg (Henry *et al.*, 1975). Prolonged increases in blood pressure also occurred in rats following daily exposure to noise, flashing lights and cage oscillation (Rosencrans *et al.*, 1966; Gamallo *et al.*, 1986), in monkeys following daily electric shock (Herd *et al.*, 1969), and in rats following daily immobilization (Lamprecht *et al.*, 1973). These experiments, which must have been particularly unpleasant for the animals, suggest that blood pressure changes could be useful in the assessment of welfare. However a complication is the finding that the dominant individuals in groups of mice or macaque monkeys had higher blood pressure than the subordinates (Ely and Henry, 1971; Shively and Kaplan, 1984) which is not so easy to interpret. In most circumstances, individuals would be affected too much by the measurement of blood pressure for the technique to be of great use in assessing welfare. However, useful information may be obtained from very tame individuals, and the preliminary results of a study using a cuff on domestic dogs are encouraging (Vincent, personal communication).

Haematocrit and haemoglobin levels change following exposure to adverse conditions for short periods but at present it is unclear how such measurements could be used as long-term indicators of welfare. Eosinophil numbers per unit volume of blood have been used by Jeppesen and Heller (1986) and Heller *et al.* (1988) as indicators of welfare in mink. Repeated immobilization of mink led to persistently high eosinophil counts. However there was initially a lowering of eosinophil counts following immobilization, particularly in female mink when their young had been removed at the customary weaning age. Obviously more information, from a variety of species, is needed before the value of this measure is clear. Other measurements of blood cell densities are mentioned in the section on immunological indicators (Section 6.6).

6.5 ADRENAL AXES

6.5.1 Medulla

Activation of the adrenal medulla is an effective response only to short-term problems. Measurements of its activity are seldom of much use as indicators when the problem is long term. However, some long-term problems involve repeated brief incidents, so assessment of the consequences of these can be useful. Habituation of the response (Rosencrans *et al.*, 1966) has to be taken into account. De Boer *et al.* (1989b) demonstrated habituation in rats to regularly repeated noise by decline in the adrenaline and noradrenaline response, but habituation to irregularly repeated noise occurred only in the adrenaline response.

Konarska, Stewart and McCarty (1989a) reported that after repeated foot-shock or restraint of rats for 27 days there was a steady decline in the increase in plasma catecholamine levels which occurred following the initial experiences, that is, there was habituation. However, the catecholamine response of these animals to a new experience, for instance a cold water swim, was higher than that produced by a cold water swim in animals which had not been subjected to prior unpleasant experiences (Konarska, Stewart and McCarty, 1989b). Hence there must have been sensitization of the adrenal medulla to such stimuli during the previous 27 unpleasant days. A comparison of adrenal medullary responses in animals with different histories can indicate those whose welfare has been poor.

Measurements of levels of adrenaline, noradrenaline, and associated enzymes and metabolites in the adrenal gland after death can provide information about the previous welfare of the animal. Noradrenaline levels in the adrenals were higher in rats which had experienced repeated immobilization (Kvetnansky, Weise and Kopin, 1970), and in mice which had experienced either repeated social confrontations (Henry *et al.*, 1971b) or repeated fights which they won (Hucklebridge, Gamal-el-Din and Brain, 1981). These authors also found higher adrenal adrenaline levels in animals which had repeated confrontations and repeated fights which they lost. The results suggest that adrenaline production is associated with fear and anxiety but noradrenaline production is associated with greater motor activity.

The mice studied by Henry *et al.* showed increased adrenal levels of tyrosine hydroxylase, which is involved in catecholamine synthesis, phenyl ethanolamine-N-methyl transferase (PNMT), which converts noradrenaline to adrenaline and monoamine oxidase, which breaks down catecholamines. Increased PNMT was also found by Hucklebridge *et al.* in frequently defeated mice, and by Mormède *et al.* (1990) in rats subjected to daily changes in social group. Tyrosine hydroxylase in the adrenal medulla increased in the study by Mormède *et al.* and in tree shrews defeated in fights (Raab and Storz, 1976). The catecholamine breakdown product vanillylmandelic acid (VMA) is higher in the saliva of people suffering from hypertension (Zielinsky, 1989), and is worth exploring further as a welfare indicator.

In general, for long-term welfare problems resulting in adrenal medulla activity, it is difficult to get useful information about live animals, and the most useful measurements are those made on the organ itself after death.

.5.2 Hypothalamic-pituitary-adrenal cortex

The first step in the activation of the hypothalamic-pituitary-adrenal cortex (HPA) axis occurs when neural activity in the hypothalamus results in the secretion first of interleukin 1ß and then corticotrophin releasing factor (CRF – also known as corticotrophin releasing hormone). As explained in Section 2.4.2, this is followed by ACTH release and glucocorticoid secretion. The extents of CRF and of ACTH production are altered by feedback from glucocorticoid levels and from later consequences of emergency system activation, such as the suppression of T-cell activity (Dantzer, personal communication). These and other modulating influences on HPA-axis activity can cause variation in hormonal responses to long-term problems.

CRF has effects in addition to initiating the production of ACTH. Prolonged problems may result in frequent bursts of glucocorticoid production in some conditions, but attenuation or suppression of production in others. There can be situation-specific variation and also individual variation in the changes in CRF production during long-term problems. Some of the mechanisms of control of HPA activity are becoming better understood as a result of the use of molecular biology techniques.

ACTH levels in plasma can be measured, but determining the significance of such measurements, as of values of glucocorticoids, is complicated by the fact that ACTH release can also be stimulated by arginine vasopressin or oxytocin (Gibbs, 1986a,b; Gaillard and Al-Damluji, 1987), and by catecholamines (Axelrod, 1984). Exposure to a continuous unpleasant situation results in an increase in plasma ACTH levels but this declines quite rapidly – after 20–40 minutes, for example, when rats were immobilized (Sakellaris and Vernikos-Danellis, 1975). Repeated stimulation, for example by electric shocks, leads to increases in ACTH initially, but these decline to basal level after 24 hours (Kant *et al.*, 1987).

(a) Glucocorticoid measures

A sufficient level of ACTH in the blood causes the release of steroids from the adrenal cortex. There will be small amounts of steroids which have effects on mineral metabolism (e.g. aldosterone) and some androgens. Principally, however, there will be a release of the glucocorticoids, cortisol and corticosterone, which affect metabolism of carbohydrates, proteins and lipids, as well as having other effects.

A single measurement of the levels of glucocorticoids in plasma, saliva or urine provides little information about the welfare of an animal over a period of more than a few hours. Even if the measurement is made in such a way that the sampling procedure itself does not upset the animal and change the level recorded, the levels will be affected by diurnal rhythms in glucocorticoid activity. Regular sampling, without disturbance, over hours or days would be required to compare the welfare of animals with different lifestyles, housing or management.

The episodic nature of glucocorticoid secretion causes difficulty in the interpretation of results. Catheterized bulls sampled hourly showed substantial variation in cortisol levels, and mean daily levels did not seem to be a useful measure of animal disturbance. However, counts of the number of peaks and of the magnitude of the peaks were of use, these being greater in tethered bulls than in free-moving bulls (Ladewig, 1984; Ladewig and Smidt, 1989).

The major reason for the inadequacy of single or occasional measures of glucocorticoid levels as welfare indicators is the adaptation of the cortex response (Sections 2.6.1 and 5.2.3). Unpleasant stimuli may cease to cause adrenal cortex responses when repeated frequently, so the value of measuring adrenocortical response declines as repetition continues. A continuing response indicates that a problem still exists, but the disappearance of glucocorticoids does not mean that the problem has been resolved. Examples of high glucocorticoid levels in prolonged difficult conditions have been described in: rats and hens at high housing density (Gamallo *et al.*, 1986; Craig, Craig and Vargas Vargas, 1986); socially subordinate mice, tree shrews and monkeys (Henry and Stephens, 1977; von Holst 1986; Sassenrath, 1970); and mice subjected to noisy, draughty conditions, in which they were also exposed to human handling and pheromones from other disturbed animals (Riley, 1981).

(b) *Adrenal cortex function tests*

Repeated exposure to different unpleasant stimuli may sensitize the hypothalamic-pituitary-adrenal cortex axis so that a test with a novel disturbing stimulus elicits a greater response than such a test would normally. After 32 days of saline injection, ether anaesthesia, noise or forced swimming, rats showed exaggerated corticosterone responses to restraint. This presumably occurred because there was greater synthetic enzyme activity, or other facilitation, in some part of the HPA axis (Restrepo and Armario, 1987)

Estimating altered synthetic enzyme activity in the adrenal cortex due to previous cortical activity is the goal of the ACTH challenge test, which involves injection of sufficient quantity of ACTH to elicit the maximum possible secretion of glucocorticoids. Using this test, Friend *et al.* (1977) found that dairy cattle had a greater plasma cortisol response after social mixing and overcrowding than they had before overcrowding. Dantzer and Mormède (1983) found that calves which had been confined in crates during rearing showed a greater cortisol response to ACTH challenge than calves reared in groups, and Meunier-Salaun *et al.* (1987) found that pigs at higher stocking densities had higher cortisol responses to ACTH challenge (Table 6.1).

Sows in group-housing showed greater cortisol responses to a standard ACTH dose and greater baseline cortisol levels if they lost aggressive encounters than if they submissively avoided encounters or won (Mendl, Zanella and Broom, 1991, 1992).

Table 6.1 Plasma cortisol level after ACTH challenge in pigs housed at three stocking densities (ng ml^{-1}) (after Meunier-Salaun *et al.*, 1987)

	Floor area (m^2) per pig			
	0.51	1.01	1.52	SE
Males	158.9	85.9	87.7	29.9
Females	107.1	58.1	90.0	12.4

SE = Standard Error

Laboratory animals exposed to prolonged stress can also show an increased response to ACTH challenge. Rats that are frequently restrained or given many high intensity shocks showed a greater corticosterone response to ACTH challenge (Sakellaris and Vernikos-Danellis, 1975; Pitman, Ottenweller and Natelson, 1990). However 18 days of restraint in a hammock resulted in no change from baseline responses to ACTH challenge in rhesus monkeys (Goncharov *et al.*, 1984). This discrepancy in results requires explanation, which may come from an examination of the responses of different individuals to various adverse circumstances.

Useful information about this mechanism could be obtained by imposing different kinds of challenge including, for example, challenges with CRF or tests of the relative suppressive effects of dexamethasone (see below). Administration of CRF would provide information about ACTH synthetic mechanisms, if carried out by intravenous injection, or about the whole set of consequences if carried out by infusion into a brain ventricle. Parrott (1990) has shown that the production of ACTH and glucocorticoids occurs without behavioural responses if CRF is given intravenously, but with the full range of the behavioural and physiological responses associated with HPA activity if CRF is infused into a brain ventricle.

The dexamethasone suppression test is used in psychiatric medicine in the diagnosis of endogenous depression and associated adrenal malfunction. Subjects with endogenous depression often show hyperactivity of the HPA axis. Dexamethasone normally suppresses ACTH production and hence glucocorticoid release, but it has no effect on those with a hyperactive HPA axis, thus suggesting malfunction of the normal negative feedback regulation of HPA activity in depressives (Hanin *et al.*, 1985). Tests on rats showed that whilst rats exposed to an electric shock from which they could escape had an HPA response suppressed by dexamethasone, those which received inescapable electric shocks had adrenal cortical responses that were not affected by dexamethasone. Sapolsky (1983) found that subordinate baboons had higher cortisol levels than dominant animals, although cortisol clearances from plasma were the same. The higher cortisol level could have arisen because the adrenal cortex was more sensitive to ACTH in the subordinate monkeys or because more ACTH was being produced. Since ACTH injection produced the same response in dominants and subordinates, but dexamethasone failed to suppress cortisol secretion in subordinates, Sapolsky concluded that the feedback regulation of the hypothalamus and anterior pituitary by glucocorticoids was relatively less effective in subordinate animals.

For all the reasons given above, direct measurements of glucocorticoid levels in body fluids after challenges to the HPA system provide limited, though useful, information. As Rushen (1991) emphasizes, however, we need to know more about how to make such challenges and how to interpret them. It is usually desirable to carry out more than one challenge, and to distinguish animals which show active responses to adversity from those which show relatively passive responses and do not use their HPA axis much.

Further tests of adrenal activity may involve looking at the size of the adrenal glands *post mortem* or studying immunosuppressive effects, many of which are caused by frequent activity of the adrenal cortex.

5 MEASURES OF IMMUNE SYSTEM FUNCTION

Animals encountering difficult conditions often show some degree of immunosuppression (Kelley, 1985; Stein, Keller and Schleifer, 1985; Calabrese, *et al.*, 1986; Calabrese, Kling and Gold, 1987; Breazile, 1988; Griffin, 1989). Much of the research documenting this point has centred on humans, but it is clear in all species that susceptibility to disease can be increased by a variety of biological disturbances. This is naturally of great importance in considerations of stress and welfare, and the nature of immunosuppression is the subject of a considerable amount of research. The aim of this section is to consider the extent to which measures of immune system function can be used as indicators of poor welfare in the long term.

Until recently, the common way to assess immune system function was to deliberately infect individuals with pathogens and determine whether or not the pathogen proliferated in the individuals, and the morbidity and mortality levels (Section 6.7). This method gives only general information about immune system responses unless it is combined with other specific measures, and it is a procedure which clearly has adverse effects on the welfare of the subjects. A refinement has been to inoculate animals with benign micro-organisms, then assess the proliferation rates and effects on the animal. Other approaches are to examine the development of tumours (a) in animals genetically prone to tumours, (b) following implantation of neoplastic cells, (c) after administration of carcinogenic substances, or (d) after inoculation of oncogenic viruses.

Using such methods, the susceptibility of animals to infection or tumour production can be compared in animals treated in different ways. For example, Riley (1975) reported that conventionally housed mice, carrying the Bittner strain of oncogenic virus, developed more mammary tumours than similar mice housed in cages in which noise, draughts and pheromones from other cages were minimized. In a similar comparison, Giraldi *et al.* (1989) found that mice in conventional cages had larger tumours and a higher rate of metastases from subcutaneous injection of Lewis lung carcinoma. Steplewski, Goldman and Vogel (1987) compared rats which were transferred to individual cages after being kept in groups with animals kept individually since weaning. After dosing with a carcinogen or inoculation with adrenocarcinoma cells, those rats which had been transferred from the group to the isolated condition had greater tumour weights. Such studies would be ethically unacceptable to most people, but they show that the environment affects the ability of an individual to resist tumour formation.

6.6.1 Measuring white cell numbers

One obvious way to discover effects of difficult conditions on an animal's ability to combat disease is to count the white cells in samples of blood or other body fluids. Changes in total white cell numbers, however, occur in various circumstances, especially when there is a pathogen attack. White cell counts in milk, for example, change when groups of cows are mixed but such results are not obtained consistently and are not easy to interpret.

A refinement of total white cell counting is to calculate ratios between one kind of white cell and another. For example, Heller and Jeppesen (1985) found that the eosinophil/lymphocyte ratio in mink was affected by holding the animals in a small wire clamp. However, longer-term problems for the animals did not have consistent effects on this ratio.

Some of the results of counting numbers of particular subsets of lymphocytes indicate that such values can be useful. Glaser *et al.* (1985) investigated levels of T-helper and T-suppressor lymphocytes in human subjects and found that these levels were reduced when the subjects took examinations. Similarly, Baker *et al.* (1985) found the lowest levels of T-helper lymphocytes in human subjects with both the highest anxiety ratings and the highest plasma cortisol levels. However, when rats were subjected to extreme treatment – up to four hours per day of escapable shock for a period of six months – there was no significant effect on the total white cell count, or the proportions of T- and β-lymphocytes (Odio *et al.*, 1986). Steplewski and Vogel's study (1986) with rats did show some effects. They immobilized rats for three hours on each of eleven consecutive days. When these rats were killed, they had higher levels of neutrophils and lower levels of lymphocytes than controls. Measurements using monoclonal antibodies showed that the immobilized rats had lower levels of total T-lymphocytes, T-helper lymphocytes and T-suppressor lymphocytes.

6.6.2 Antibody production

Antibodies are immunoglobulins of four types known as IgA, IgE, IgG and IgM which are produced by ß-lymphocytes. Total quantities of immunoglobulins can be measured in plasma or colostrum, and IgA can be measured in saliva. Total immunoglobulin levels have been found to be of little use in welfare assessment, although eight days after injecting tethered and group-housed sows with *Escherichia coli* K99, Barnett *et al.* (1987) found lower IgG and IgM levels, correlated with higher plasma glucocorticoid levels, in the tethered sows than in the group-housed sows.

The production of specific antibodies following an experimental challenge with an antigen can give valuable information about welfare. Animals can be detrimentally affected by a housing system or treatment in such a way that they are less able to produce antibodies following antigen challenge. Metz and Oosterlee (1981) found that the antibody response to the injection of sheep's red blood cells was less in recently tethered than in untethered sows. In a similar experiment, antibody production following injection of the bacteriophage X174 was considerably depressed in six-month-old squirrel monkeys separated from their mothers compared with those not separated from their mothers (Coe, Rosenberg and Levine, 1988).

In recent work by Zanella, Broom and Mendl (1991) involving challenging sows with vaccines, sows given tetanus toxoid or atrophic rhinitis vaccine produced antibodies to these in blood and colostrum a few days later. As shown in Fig. 6.2, sows which showed a greater response to ACTH challenge produced lower levels of antibodies following antigen challenge, suggesting that activation of the hypothalamic-pituitary system was suppressing antibody response.

Other tests of the ability of animals to produce antibodies when challenged by an antigen are the haemagglutination and plaque-forming cell assays. Haemagglutination occurs when serum taken from an individual which has had foreign red blood cells injected into it five days earlier is added to those red blood cells; serum from immunosuppressed animals produces less haemagglutination.

The plaque-forming cell assay (Esterling and Rabin, 1987) is carried out with lymphocytes from the spleen of an animal previously sensitized to foreign red blood cells. Spleen samples would normally be taken from an animal after death and plaques allowed to form when the lymphocytes

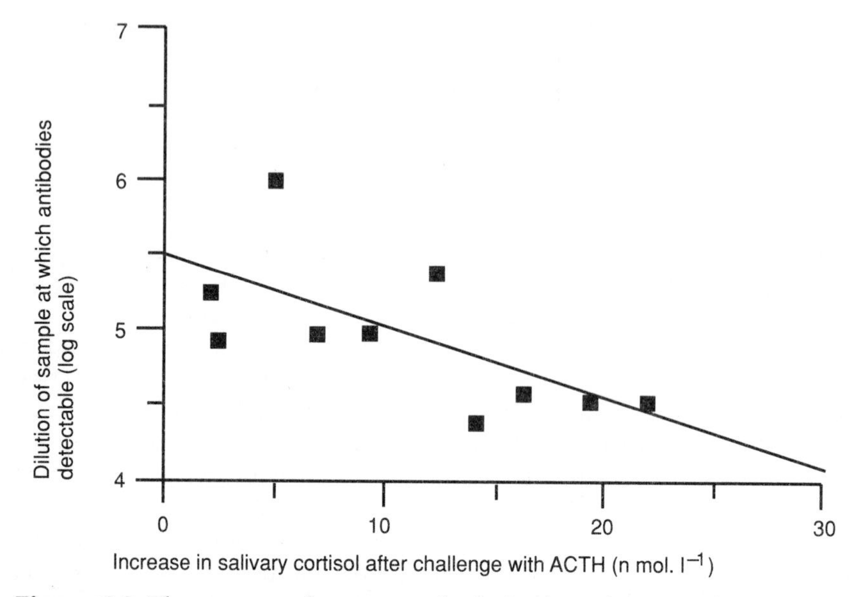

Figure 6.2 The amount of tetanus antibody in the colostrum of sows after a challenge with the tetanus toxoid is measured by the extent to which the sample can be diluted and still show detectable antibody. Sows which showed a smaller cortisol response to ACTH challenge had higher antibody levels (from Zanella, Broom and Mendl 1991; Zanella, 1992)

are incubated with the red blood cells in the presence of complement. Rabin *et al.* (1987) found that male mice housed five to a cage showed a lower plaque-forming response than those kept individually, and hence suggested that there was an association between the increased fighting which occurs in male mice crowded in a cage and reduced T-cell function. Females, which usually do not fight, did not show this difference.

6.6.3 T-lymphocyte function

Antibody production is affected by the lymphokines produced by one of the forms of T-lymphocyte, the T-helper cells, so measurement of reduction in antibody levels is an indirect way of ascertaining the difficulty an animal has in coping with its environment. T-lymphocytes can act directly on foreign antigens and their action is affected by an animal's success in coping, so measurement of T-cell activity can be a useful indicator of how poor welfare is. However, interpretation of such measurements requires care, as T-cell proliferation and activity can be induced by various factors, especially pathogen presence. Skin sensitivity tests have been used for many years in tests of immune system responses. The thickening of the skin which follows application of substances such as 2,4-dinitro-1-fluorobenzene is mediated by T-lymphocytes. An inflammatory response is also caused by intradermal injection of the plant lectin phytohaemagglutinin (PHA). The delayed hypersensitivity test also depends on such a response; it is usually carried out by injecting an animal, such as a mouse, with sheep's red blood cells and assessing the skin thickening response 25–72 hours after further sheep's red blood cells are injected into the foot pad. Studies which have utilized such tests include those of Kelley *et al.* (1982) who found that contact sensitivity to dinitro-fluorobenzene and delayed hypersensitivity responses were reduced in calves which had been kept at –5 °C or 35 °C as compared with calves kept within a neutral temperature range. Tests involving injection of the foot pad are very likely to cause significant discomfort (Barclay, Herbert and Poole, 1988) and the results of any test using a non-specific lymphocyte stimulator like PHA will be affected by each subject's previous experience of antigens, so spurious results can be obtained (Blaese *et al.*, 1973). Because there is a limit to the precision and range of measurement of a skin thickening response, these tests can be used only as a rough indicator in an assessment of long-term welfare problems.

A more precise experimental method of assessing T-cell activity involves the use of mitogens with cultures of lymphocytes, and an assessment of their rate of proliferation by measuring the rate at which radio-labelled thymidine is taken up. The three mitogens which are most frequently used are concanavalin A (ConA), phytohaemagglutin (PHA) and pokeweed mitogen. Lymphocytes from blood or the spleen can be used, and it is clear from work such as that of Barta (1983) or Murata (1989) that there are factors in the serum of animals, after they have been exposed to difficult conditions, which inhibit lymphocyte cell division.

The amounts of such serum factors are affected by a wide variety of treatments. When Monjan and Collector (1977) subjected mice to high noise levels, plasma corticosterone increased, and the rate of division of splenic lymphocytes in the presence of ConA was reduced in comparison to mice not subjected to noise. As the mice habituated to the noise the suppression of T-cell proliferation declined. Rats exposed to inescapable shock for six months also showed a reduced lymphocyte proliferation response to ConA and PHA in proportion to the amount of time per day that the shock treatment continued. Crowded male mice have been shown to have a smaller lymphocyte proliferation response to ConA and PHA than singly housed male mice (Rabin *et al.*, 1987).

A variety of social deprivation studies have revealed effects on lymphocyte proliferation. Mother and infant bonnet macaque monkeys had reduced mitogenesis with PHA and ConA after they were separated, but mitogenesis returned to normal after they were reunited (Laudenslager,

Reite and Harbeck, 1982). Also, rhesus monkeys which were frequently separated from their mothers when between three and seven months of age showed depressed mitogenesis to PHA for more than a year after the separations occurred (Coe *et al.*, 1989). Separation of pig-tailed monkeys which had lived together since leaving their mothers also caused such effects. Separation resulted in agitated behaviour and vocalization as well as substantially reduced mitogenesis with ConA and PHA for several weeks. Depressed human patients (Stein, Keller and Schleifer, 1985; Calabrese *et al.*, 1986) and people who had been bereaved (Bartrop *et al.*, 1977; Schleifer *et al.*, 1983) showed reduced mitogenesis with PHA and ConA as compared with matched controls or with measurements of the same people at other times. Levels of mitogenic suppression have proved to be more sensitive indices of welfare than total lymphocyte counts, proportions of different lymphocytes, or delayed hypersensitivity.

One particular kind of lymphocyte whose action is of considerable importance in body protection and whose cytotoxic action can be assessed in an assay are the natural killer (NK) cells. The NK cell assay described by Ullberg and Jondal (1981) involves culturing the lymphocytes with chronic myeloid leukaemia cells (K-562) labelled with radioactive 51Cr, which is released when the target cells are killed. An assay using mouse lymphoma cells (YAGI) involves incubation with lymphocytes previously made fluorescent so that binding can be assessed using ultraviolet spectroscopy. The NK cell activity of mouse spleen cells was reduced after prolonged exposure of the animals to noise (Monjan and Collector, 1977) or limb amputation (Pollock *et al.*, 1987). Human subjects showing depressive symptoms after bereavement had blood lymphocytes with lower NK cell activity than people who had not been bereaved (Irwin *et al.*, 1987). Medical students also showed reduced NK cell activity before and during examinations (Glaser *et al.*, 1986b).

Another way of assessing T-lymphocyte activity is to measure the production of lymphokines such as interleukin-2, which can be produced by lymphocytes and can cause proliferation of T-helper cells, and interferons which are antiviral agents produced by T-lymphocytes. Glaser *et al.* (1986b) found that interferon production by human lymphocytes was lower in students at examination time than at other times. In the study by Rabin *et al.* (1987) of male mice crowded together, the cell-free supernatant from their spleen produced less interleukin-2 than that of individually housed mice.

6.6.4 Other body defences

Foreign particles and cells which get into the body are removed by the phagocytic activity of neutrophils or monocytes. The effects of environmental conditions on animals have been assessed by measuring the ability of peritoneal macrophages from mice to kill leukaemia cells *in vitro* (Pavlidis and Chirigos, 1980), and the ability of macrophages to ingest bacteria or yeasts (Blecha, Boyles and Riley, 1984) or other particles (Okimura *et al.*, 1986). The functioning of the complement system, which facilitates the breakdown of foreign cells (Coe, Rosenberg and Levine, 1988), and the operation of the blood platelet system can also be impaired in an individual that is having difficulty in coping with its environment.

.7 DISEASE INCIDENCE MEASURES

An enormous body of literature exists on the interrelations between environmental conditions and the consequences of infection by pathogens (for instance, in fish, Pickering, 1989b). As emphasized earlier, the welfare of diseased animals is poorer than that of non-diseased animals, the extent of the effects on welfare depending on how much the animal has to do to combat the disease, how great is the body damage caused by the pathogen, and how much the animal suffers because of the disease. Pasteur reported that chickens whose legs were immersed in cold water became more susceptible to anthrax (Nichol, 1974). Of the many studies carried out since then, few have been concerned with trying to assess the welfare of animals but many are in fact relevant to this aim.

Changes in disease incidence with animal husbandry have been reported. For example, Sainsbury (1974) states that a gradual increase in chronic infections of poultry occurred when the incidence of intensive production practices was increased. Many factors are involved in the causation of environmentally evoked diseases (Ekesbo, 1981), and care must be taken when trying to relate an increase in disease incidence to any one factor. For example, a higher level of a disease in one kind of cattle housing system as compared with another could be a direct consequence of the system itself, or could be related to management practices.

An illustration of the relation between disease susceptibility and difficulty in coping with environmental conditions comes from studies on chickens. When birds which were strangers were put together they displayed, fought and had increased adrenal cortex activity. After such social mixing had occurred, challenge with *Mycoplasma gallisepticum*, Newcastle disease, Marek's disease or haemorrhagic enteritis resulted in greater pathogen levels in the body, greater morbidity and greater mortality than in chickens that were not mixed with strangers (Gross, 1962; Gross and Colmano, 1965; Gross and Siegel, 1981). When challenged with *Escherischia coli* or *Staphylococcus aureus*, however, social mixing led to increased resistance (Gross and Colmano, 1965; Gross and Siegel, 1981). The causes of this apparently anomalous result were revealed by measuring antibody levels. Social mixing was found to reduce antibody activity against both viral antigens, such as those causing Marek's disease, and particulate antigens such as those from *E. coli* (Gross and Siegel, 1975; Thompson *et al.*, 1980). This immunosuppression was probably mediated via increased corticosterone levels and reduced interleukin II production by T-cells, in the socially mixed chickens. However, another effect of glucocorticoids such as corticosterone can be to counteract inflammatory responses like those caused by *E. coli* and *S. aureus* so the effects of these pathogens are ameliorated following the social mixing (Siegel, 1987). In the *E. coli* challenge experiment, the anti-inflammatory response of the elevated corticosterone was more important than the immunosuppressive response. In most situations, immunosuppression resulting from increased adrenal cortex activity is of the greatest importance.

6.8 OPIOIDS

Endogenous opioids such as endorphins, enkephalins and dynorphin are involved in many different body control mechanisms in the brain (Akil, Watson and Young, 1984; Akil, Shiomi and Matthews, 1985; Smelik, 1987) including the release of prolactin, luteinizing hormone and growth hormone (Grossman and Rees, 1983), and in pain perception and reward motivation. ß-endorphin is also secreted into the blood in parallel with ACTH from their mutual precursor pro-opiomelanocortin (Guillemin *et al.*, 1977; Rossier *et al.*, 1977). Hence treatments which elevate ACTH and glucocorticoid production may also elevate plasma levels of ß-endorphin. Plasma ß-endorphin increases in circumstances such as: surgery in humans (Cohen *et al.*, 1981; Dubois *et al.*, 1981; Smith, Besser and Rees, 1985); mulesing, castration and tail-docking operations in sheep (Shutt *et al.*, 1987); anaesthesia and surgery in horses (Taylor, 1987); shearing and electro-immobilization of sheep (Jephcott *et al.*, 1987); handling, transport and slaughter of sheep (Fordham *et al.*, 1989); stunning of sheep (Anil *et al.*, 1990); and isolation of sheep from their flockmates (Al-Gahtani and Rodway, 1991). The ß-endorphin could serve to transmit information to peripheral organs, or it could have some analgesic function, or it could do both. However in the study by Smith, Besser and Rees (1985) the ß-endorphin level was not related to reported pain, and in the study by Shutt *et al.*, behavioural signs of discomfort in sheep persisted after ß-endorphin levels had dropped.

The analgesic action of endogenous opioid peptides has been known since the work of Hughes *et al.* (1975). Their action can explain the reports by people who lose a limb in battle or who are badly injured in sporting competitions that they were unaware of the injury until some time after it occurred. The analgesia in such circumstances must be very rapid and effective. Opioid production in difficult conditions is not limited to those in which painful injuries occur but is a much more wide-ranging response. For example, plasma ß-endorphin levels increased during restraint in rats (de Souza and van Loon, 1985) and in rhesus monkeys (Kalin *et al.*, 1985a) but, in the latter study, levels of ß-endorphin in cerebro-spinal fluid were not affected. In male talapoin monkeys kept in groups, the level of ß-endorphin in the cerebrospinal fluid was highest in subordinate individuals, lower in those of intermediate rank and lowest in dominant individuals. Hence opioids may have a role where there are long-term welfare problems.

The possibility that some animals may self-narcotize as a means of trying to cope with prolonged difficult conditions such as close confinement has been suggested by Cronin, Wiepkema and van Ree (1985) and Broom (1986b, 1987). This raises a question of interpretation: that of deciding how self-narcotization using endogenous opioids affects welfare. As discussed in Chapter 4, the welfare of an individual which has to do a lot in order to cope (for instance, to use endogenous opioids) is worse than that of another individual which does not have to do this. If an individual is in pain, however, its welfare is worse than that of another individual which blocks the pain with an endogenous opioid. Hence the welfare scale that is operating is similar to that shown in Fig. 4.6.

The role of endogenous opioids in animals encountering difficult conditions has sometimes been studied by the use of endogenous opioid antagonists which block the receptors in the brain to particular opioids. The relevant brain receptors are mu (μ) receptors for endorphin, delta (d) receptors for enkephalins and kappa (k) receptors for dynorphin. Cronin, Wiepkema and van Ree (1985) and Cronin *et al.* (1986) used naloxone, which is particularly effective at blocking μ receptors, in a study of tethered sows with high levels of stereotypies such as bar biting. When such animals were injected with naloxone, the occurrence of stereotypies was reduced during the short period when naloxone acts, hence Cronin *et al.* proposed that stereotypies result in the production of ß-endorphin and self-narcotization. However, at the doses used, μ,d and k receptors would be blocked temporarily so many systems would be affected. Also, Cohen *et al.* (1983) reported that naloxone injection caused human subjects to feel bewilderment, anxiety, headache, dizziness, nausea and stomach ache. Hence the naloxone effect on stereotypy occurrence could be due to one of the side-effects or, if affecting opioid action, could affect any one of the different types of opioids.

Other evidence exists for links between behavioural responses in difficult situations and opioid action. Hiramatsu *et al.* (1987) found that administration of opioid agonists in rats increased circling behaviour. Similarly, Cancela, Artinián and Fulginiti (1988) found that injection of naloxone did not affect stereotypy incidence in rats. Stereotypies and injurious behaviour in mentally retarded humans, however, were associated with higher plasma ß-endorphin levels (Sandman *et al.*, 1990). A further link between opioids and stereotypies was provided by the results of a study by Friederich, Friederich and Walker (1987) who injected opioid and non-opioid fragments of dynorphin unilaterally into the nigral region of the brains of rats. This induced circling behaviour in the rats which were not inhibited by the endogenous opioid antagonists naloxone, WIN 44, or 441-3. Hence it may be that kappa receptors have some link with stereotypies and non-opioid mechanisms are also likely to be involved.

Opioid receptors in the brain are not fixed in number: their frequency of occurrence can be altered by various factors, including social isolation, electric shock to the feet, restraint, or the use of agonists or antagonists (Bonnet, Hiller and Simon, 1976; Nabeshima *et al.*, 1985; Zeman, Alexandrova and Kvetnansky, 1988; Blanchard and Chang, 1988). Nabeshima *et al.* found that a brief electric shock decreased d agonist binding and increased μ antagonist (naloxone) binding in rats. In the work by Bonnet, Hiller and Simon, the effects of social isolation on the binding capacity of opioid receptors in mice was positively correlated with behavioural evidence for analgesic responses to morphine.

Long-term housing conditions and the behavioural responses they induce have been related to opioid receptor density by Zanella, Broom and Hunter (1991, 1992). Sows which had been tethered for some months had a higher μ receptor maximum binding capacity in the frontal cortex than did group-housed sows. Tethered sows which were inactive for much of the time had high μ receptor densities, whereas those which showed high levels of stereotypies had lower k receptor densities. If lower receptor densities indicate down-regulation, i.e. a reduction in receptor density resulting from earlier high levels of the opioids binding to them, these results suggest that stereotypies may be related to high levels of dynorphin, and inactivity may be related to low levels of endorphin.

There is some difficulty in interpreting the results of work on opioid peptides. It seems certain that some changes in opioid levels and action are related to welfare. However we do not yet know which changes are a direct adaptive response to some difficulty which the animal encounters, which changes occur at the same time as ACTH production but are unrelated in function, which changes have an analgesic effect, which changes are a consequence of some behavioural response, or which changes alter the probability that some behavioural response is shown. It is likely that several different measures of opioid activity will help in the assessment of welfare, but it is not yet clear which measures will be useful and how information about opioids will be related to that from other measurements of welfare.

6.9 BEHAVIOURAL MEASURES

The best indicators of long-term problems for an animal are frequently measurements of behaviour. One approach for identifying conditions which constitute poor welfare is to observe an animal's behaviour when it is faced with them; such tests of preference and aversion are discussed in Chapter 7. Similar tests, of direct avoidance by animals of conditions in which they have recently been kept, may also be made, though these indicate short-term rather than long-term welfare problems.

Various other analyses of behaviour also provide information about an animal's welfare. The simplest occurs when an individual has difficulty carrying out normal movements. A second is associated with the lack of a resource, or some specific frustration. A third group of behaviours arises as a consequence of frustration, inability to escape from perceived danger or unpleasant stimulation, an overall lack of stimulation, or confusion when too much is happening for the animal to make appropriate responses. A feature of all these difficult situations is the animal's lack of control of its interactions with its environment.

On many occasions behaviours which indicate an animal's welfare are part of its attempt to cope with an environmental difficulty. However, for some of these behaviours there is no evidence that the behaviour is helpful. It may in fact be making the situation worse for itself or for other animals; this is a behavioural pathology. In either case, measures of such behaviour indicate welfare problems, since welfare is poor both when an animal is having difficulty in its attempts to cope and when it is failing to cope.

9.1 Problems with movement

Movement difficulties

It is obvious that an individual which cannot walk has difficulties in coping with its environment. If the individual is capable of walking but, because the floor is slippery, it does not do so, or if it does not do so in a normal way, the same must be said. This situation does arise sometimes on farms, and Andreae and collaborators have described the behaviour of cattle kept on slippery slatted floors (Andreae, 1979; Andreae and Smidt, 1982).

The normal movements of cattle standing or lying are shown in Fig. 6.3.

When the animals were kept on slippery floors, they differed in the time, sequence and patterns of these movements. The first stage in lying down usually involved lowering the head and apparently sniffing the ground. After a few seconds, this was followed by the rest of the lying sequence on non-slippery floors, but only after prolonged pauses if the floor was slippery. Other interruptions in the sequence also occurred (Fig. 6.4), so the whole process of lying could take as long as 20 minutes.

Sometimes the animal even changed the order of movements completely and lay down rump first (Fig 6.5). It is easy for a person to

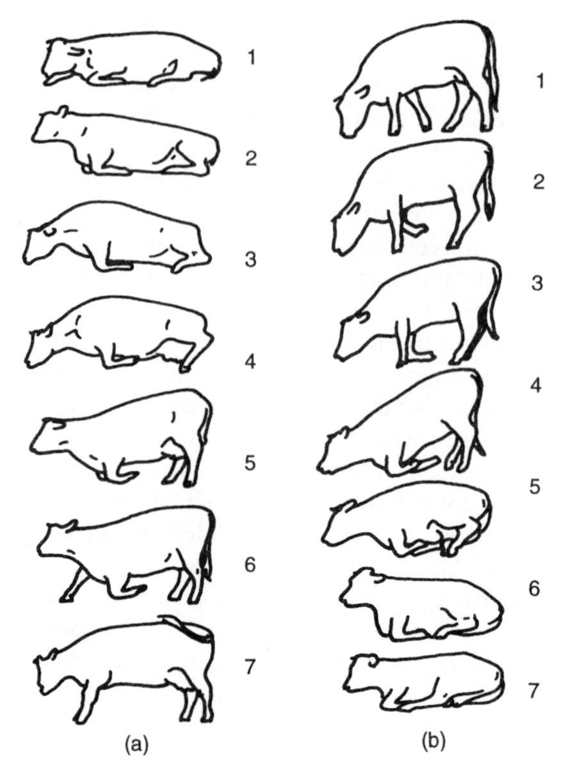

Figure 6.3 The typical sequence of movements which occurs when a cow stands up (a) and lies down (b) (from Andreae and Schmidt, 1982).

imagine these abnormalities of behaviour after trying to stand or crouch on ice.

Animals which have lived in conditions that preclude exercise may have movement difficulties. The veal calf which has lived for 4–6 months in a narrow crate, or an animal that has been kept on a short tether may find walking and other locomotion difficult (Trunkfield, 1990). A bird kept in a cage too small for wing movements is likely to be inefficient at flying. Direct effects of the confinement on the muscles, bones and perhaps nervous control mechanisms can account for these behavioural abnormalities.

(b) Movement prevention

While an animal is closely confined it will be unable to carry out certain movements, and modifications in behaviour will occur as a consequence. Other changes in behaviour may also become evident, owing perhaps to

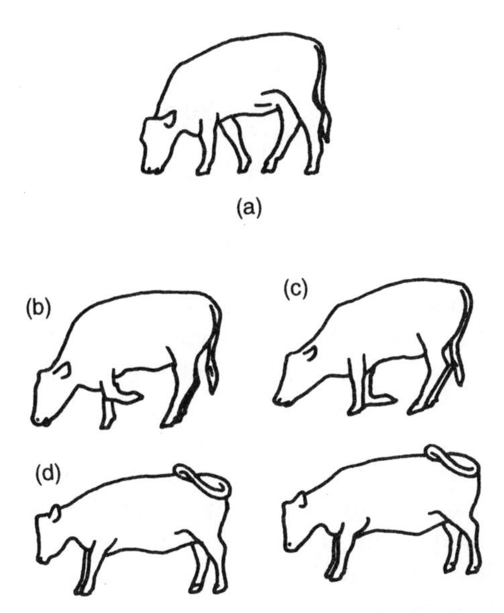

Figure 6.4 Young cattle on slippery slatted floors show alterations in behaviour which inhibit, delay or prolong lying. (a) Repeated ground sniffing without lying down; (b) leg bent in without floor contact; (c) leg bent in with floor contact; (d) lying down interruptions. For the normal lying sequence, see Figure 6.3 (from Andreae and Smidt, 1982).

Figure 6.5 In the same situation as that in Figure 6.4, some young cattle lie down rump first, presumably to minimize painful events when trying to lie on the slippery floor (see also Figure 6.3) (from Andreae and Smidt, 1982).

frustration. Sows which are kept in stalls and veal calves kept in crates may have difficulty in lying down because of the cramped conditions. Normal lying movements may even be completely prevented. Some individuals respond to this by 'dog-sitting' in which the hind quarters are on the ground but the front legs are extended, a posture that is rare in more spacious conditions.

Confinement of circus animals, zoo animals, pets and some farm animals prevents them executing their normal range of locomotor, grooming, food-finding, sexual and other socially orientated movements. The consequence may be grossly modified attempts to show these movements. The results may be stereotypies, but need not be; in some cases, parts of normal behavioural sequences may be shown. Inability to complete normal grooming sequences frequently results in excessive grooming of the body parts which can be reached, for example, in veal calves which cannot turn around to groom their hindquarters. Inability to carry out sexual movements may result in bizarre sequences of substitute movements including elements which appear to be part of sexual display or attempts at sexual stimulation.

6.9.2 Behaviour associated with lack of a resource

Animals which are deprived of part of their nutrient requirement, or a specific component of it, may show characteristic behaviour responses. Those responses often include movements associated with finding or obtaining the food. Carnivores might carry out prey-catching movements, and animals which find food items by sifting through earth or litter may carry out movements with a similar effect. Animals deprived of a component of their diet will often eat or chew a variety of materials which they would not otherwise eat. Phosphorus deficiency in farm animals can lead to 'pica', which is the chewing of wood, bones, soil, and so on; other abnormal feeding behaviour may also be the result of some dietary deficiency (Fraser and Broom, 1990). The eating of such materials, or of hair or faeces, must be considered an indication of a welfare problem. In the examples given, the behaviour is related to the acquisition of the resource that is lacking. Other behaviour in such situations may be initiated by the deficiency, but some of the responses can seem unrelated to the problem.

Sows confined in a stall often show bar-biting, drinker-pressing and sham-chewing stereotypies (Jensen, 1979, 1980; Rushen, 1984; Blackshaw and McVeigh, 1984; Broom and Potter, 1984). These may indicate a general frustration but it has been suggested that the movements are a specific response to low food availability. Appleby and Lawrence (1987) point out that sows fed only 2 kg of food would eat more than 5 kg if given the opportunity, and that most of the stereotypies shown by confined sows incorporate movements which are a part of feeding movements. Well-fed sows show lower frequencies of stereotypies (Robert *et al.*, 1992). However, some of these movements could also have other functions. In the study by Broom and Potter (1984) provision of an extra 2 kg of virtually non-nutritious oat hulls for sows increased lying time and the stereotyped sham chewing which occurs whilst lying, but did not change the total duration of stereotypies. It may be that the lack of nutrients is the cause of this stereotypy and increased bulk of food in the gut is not sufficient to prevent it, but it seems more likely that the causation of the abnormal behaviour is multifactorial, only one of the factors being related to food.

.9.3 Behaviour associated with lack of social or sexual partners

Searching for conspecifics is adaptive in social animals because the presence of companions may help in predator avoidance, food finding, food acquisition and environmental control (Broom, 1981b), in addition to providing potential sexual partners. The presence of conspecifics enriches an individual's environment and opens up possibilities for a wide range of different experiences and behaviours. If animals normally kept on farms – virtually all of which are social species – are isolated, they show a variety of behavioural abnormalities which could be due to a lack of social stimu-lation. There is certainly a great reduction in the occurrence of abnormal behaviours when social companions are present.

Individually housed calves show much self-licking and repetitive tongue rolling. The incidence of these behaviours is usually less in socially reared calves, perhaps because of the possibilities for social interaction but probably also because of the opportunities provided by greater space for movement. Stereotypies are also much rarer in group-housed sows or horses than in individually housed animals. High levels of aggression have been reported in rodents housed in isolation for long periods (e.g. Goldsmith, Brain and Benton, 1976) although very prolonged isolation from an early age can result in low levels of aggression because behaviour is affected to the extent that co-ordinated aggressive behaviour is not possible (Cairns, Hood and Midlam, 1985). Tethered sows have also been reported to be particularly aggressive (Barnett *et al.*, 1987). However, prolonged isolation may lead to reduced reactivity to novel situations, reduced interaction with inanimate stimuli, poor social responsiveness, and failure in social competition in lambs, calves and foals (Zito, Wilson and Graves, 1977; Broom and Leaver, 1977, 1978; Broom, 1982; Houpt and Hintz, 1983).

The common practice of rearing domestic animals in isolation leads to a variety of abnormalities of sexual behaviour (Beilharz, 1985; Price, 1985a). Bulls and goats can show disorientation during copulation if they have had little social experience, but juvenile mounting experience reduced the incidence of this in beef bulls (Silver and Price, 1986). Rams needed contact with females during adolescence to show normal copulatory behaviour (Orgeur and Signoret, 1984). Isolation-reared boars showed various inadequacies in mating behaviour, but contact with other boars through wire mesh reduced these (Hemsworth *et al.*, 1978; Hemsworth and Beilharz, 1979).

Reduced social experience early in life can lead to rejection of the young or to poor parental behaviour in primates (Harlow and Harlow, 1965; Chamove, Rosenblum and Harlow, 1973), cattle (Broom and Leaver, 1977; Broom, 1982) and horses (Houpt, 1984; Houpt and Olm, 1984).

6.9.4 Consequences of inability to perform a behaviour

Though much of the behaviour of animals is clearly oriented towards acquiring resources, in certain circumstances individuals apparently try hard to carry out a behaviour which is normally the means of achieving an objective. The behaviour itself is apparently an objective since, in addition to making considerable efforts to show the behaviour, animals prevented from doing so show behavioural and physiological abnormalities. Examples of functional systems in which this occurs include keeping the body clean, having adequate knowledge of the environment, and preparing for reproduction or predator avoidance by establishing social relationships. The behaviours for achieving these include, respectively, grooming or preening, exploration or curiosity, and social interactions. Animals make considerable efforts to show these behaviours and exhibit clear signs of disturbance if they are unable to do so. As a consequence of the fact that some biological needs (Chapter 4) can be remedied only by being able to show certain behaviours, there are various behavioural and other consequences of living where such essential behaviours cannot be shown.

One behaviour which seems to be important in itself is suckling by young mammals. This is defined as the obtaining of milk from the mother or another female by sucking a teat. There is an obvious nutritional objective, but the sucking behaviour itself also seems to be necessary. Young calves will drink as much milk from a bucket as they would take when suckling their mother, then spend long periods afterwards sucking the bucket handle, the bars of their pen or projecting parts of the anatomy of other calves. The desire to show such sucking behaviour on a non-nutritive teat declines with age (Fig. 6.6; Broom, 1991c; Broom and van Praag, in preparation). Human babies drink rapidly from a bottle but will also suck at fingers, appropriately shaped toys or pacifying dummies.

Piglets will continue to suckle until 8–14 weeks if not separated from the mother (Jensen, 1986). Hence it is not surprising that they show

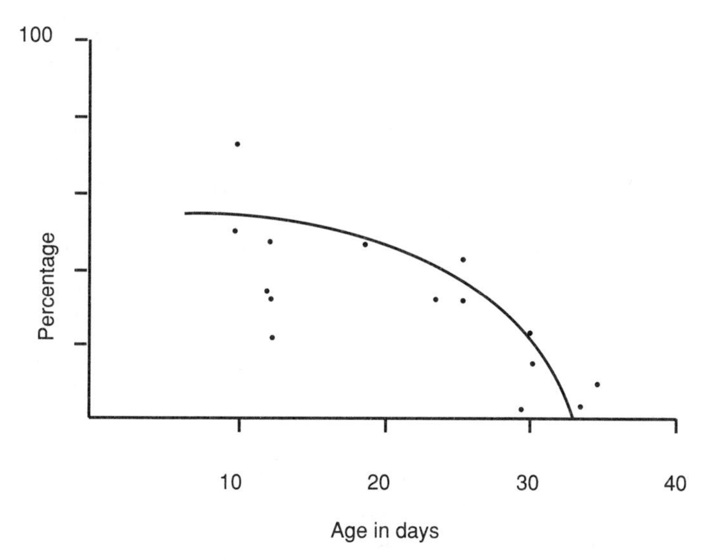

Figure 6.6 Percentage of time sucking or licking a novel non-nutritive teat or another calf during 10 minutes after drinking milk from a bucket (Broom and van Praag, in preparation).

belly-nosing behaviour to other piglets if weaned at 3–5 weeks. Belly-nosing is a movement of the snout on the belly or soft tissue between the hind or forelegs of another piglet. As pointed out by van Putten and Dammers (1976) and by Schmidt (1982) it is similar to the massaging movements which piglets direct towards the udder of the sow. Belly-nosing is often followed by sucking on the penis or navel. The frequency of belly-nosing can be reduced by the provision of straw to chew (van Putten and Dammers, 1976; Schouten, 1986).

The manipulation of material such as straw is an activity which seems to be important for animals of several species, whilst rooting in soil or straw is a favoured activity of pigs. Rooting by domestic pigs presumably derives from the search for food in wild pigs, and the manipulation of the stems and leaves of vegetation is a part of natural feeding behaviour in cattle. When these activities are not possible, animals which are well fed still show abnormalities of behaviour. In such conditions, pigs, hens and calves may show stereotypies, and calves may show excessive licking. The incidence of such behaviour can be reduced in all these species by providing straw and, in pigs, by providing earth in which to root (Fraser, 1975).

Pigs living in groups without adequate environmental stimulation develop a behaviour in which the tails of other pigs are first manipulated and then bitten. This behaviour is less frequent if manipulable material is available for the pigs (Fraser and Broom, 1990). The environments in which animals show sterotypies, such as excessive licking or tail biting are usually ones in which there is little for the individual to do. The behavioural abnormalities may be a consequence of living in a barren environment in which there is little stimulation and a lack of opportunity to show various behaviours, but it is clear that one of the forms of environmental enrichment which reduces the incidence of abnormal behaviour is the provision of manipulable material. The general appreciation that welfare problems arise in many species of animals when they live in barren conditions has led to increasing research on environmental enrichment (Chapter 7).

6.10 OTHER CONSEQUENCES OF FRUSTRATION AND LACK OF CONTROL

As explained in Chapter 4 and mentioned earlier in this chapter, many situations which result in poor welfare are those in which an animal lacks control over its interactions with the environment. Such situations often involve frustration or unpredictability. When a resource is lacking or a behaviour cannot be carried out, one part of the problem for the animal is the direct effect of the lack of a resource, such as water, and another is the feeling of frustration because the particular regulatory system cannot work properly. Similarly, an animal which might be subjected at any time to an attack by a conspecific will be disturbed both by the attack and by the uncertainty about when the unpleasant event will occur. Behavioural responses in such situations include increased aggression, stereotypies and apathetic or unresponsive behaviour.

6.10.1 Aggression

Aggressive acts in frustrating situations are well known to us all. They can occur in domestic or laboratory animals which are subjected to frustration about, for example, access to food. In the experimental situation described in Chapter 3 (Duncan and Wood-Gush, 1971) in which two hens were put into a cage in which they expected to be fed, but their feeding was thwarted by a perspex cover over the food container, one of the results was in increase in aggression. One hen pecked the other, even though she was not the cause of the frustrating situation, and the extent of pecking was altered by the duration of food deprivation (Table 3.2).

An example of aggression arising due to unpredictability is the experimental study on pigs by Carlstead (1986) (Section 3.7.2). In these examples, the aggression is an indicator that the welfare of the aggressor is poor, though aggression does not necessarily indicate poor welfare of the aggressor in all situations. Most aggression does, of course, also adversely affect the welfare of the individual which is the target of the aggression.

.10.2 Stereotypies

Repetitive and stereotyped behaviours are amongst the most important indicators of long-term welfare problems. The best known examples are the route-tracing of human prisoners and of animals kept in cages, for example in zoos. Hediger (1934, 1941) and Meyer-Holzapfel (1968) describe route tracing and other stereotypies in zoo animals while Keiper (1970) studied the paths followed by canaries in cages. Scientific studies of human prisoners are not easy to find, but Charrière (1969) in his novel about convicts in prisons in the French West Indies gives vivid accounts of the repeated movements of those kept in solitary confinement for long periods. The rocking and weaving movements of children with autism or other psychiatric disorders are well known (Levy, 1944; Hutt and Hutt, 1970); Levy also describes various stereotypies in other species. Crib biting and wind-sucking by horses are described by Brion (1964), bar biting by sows is reported by Fraser (1975), and there are reviews of stereotypies by Ödberg (1978), Broom (1981, 1983b), Dantzer (1986), Fraser and Broom (1990) and Mason (1991a, in press).

Detailed descriptions of stereotypies (Stolba, Baker and Wood-Gush, 1983; Fraser and Broom, 1990) have demonstrated that there is some, but not much, variation in the sequence of movements displayed. Movements with an obvious function, such as rumination, locomotion and some displays, are not referred to as stereotypies. Hence the definition of a stereotypy is: a repeated, relatively invariate sequence of movements with no obvious function. Some of the movements which are repeated are brief, as in sham chewing by sows or rocking by children; others, like route-tracing sequences by bears in zoos, are lengthy. Some movements are repeated regularly whilst others are not. People in situations where they lack control temporarily often show intermittent stereotypies, for example key jangling or pacing by an expectant father in a hospital waiting room.

The question of how to decide whether an apparent stereotypy has a function is usually quite easily answered; whilst a single movement may be part of a normal functional system, frequent repetitions of movements are necessary only for certain limited purposes. These purposes include locomotion to a particular place, and repeated feeding, respiratory, cleaning or display movements. A brief period of observation is usually sufficient to distinguish stereotypies from such movements. In the case of some stereotypies the movement is so bizarre that it is clearly unlikely to have a function, for example Mason (1991c) describes a female mink in a 75×37.5×30 cm mink farm cage which would repeatedly rear up, cling to the cage ceiling with her front paws and then crash down on to her back. Some other stereotypies, however, are sufficiently similar to normal behaviour, or have sufficient possibilities for real function, to create doubt about whether or not they are stereotypies. Careful study usually clarifies the situation but there remain questions about whether behaviours such as thumb sucking and repetitive play by children, or wheel running in captive rodents should be classified as stereotypies. In the first example some improvement in motor and cognitive development may result (Bateson, 1986) and in the second the animal gets necessary exercise from the movement. Similarly, it might be argued that repeated tongue-rolling and sucking by early-weaned veal calves has a function, because calves need to suck something and the tongue sucking helps to satisfy that need. However, in these and other instances, the activity is not effective in helping the development of the young animal, or in providing exercise or satisfying a need. In addition, none of the possible beneficial effects of such movements require that the action has to be repeated in almost exactly the same way each time it occurs. Hence, in many cases, the repetition of the relatively invariate sequence of movements does not have a function and the behaviour is a stereotypy. Factors affecting the variability of stereotypies and questions about the development of stereotypies are reviewed by Mason (in press) and Mason and Turner (in press).

As discussed by Dantzer (1986), Mason (1991b) and others, in most cases we do not know whether a stereotypy is helping the individual to cope with the conditions, has helped in the past but is no longer doing so, or has never helped and has always been a behavioural pathology. But in all cases the stereotypy indicates that the individual has some difficulty in coping with the conditions, so it is an indicator of poor welfare. Some stereotypies must indicate worse welfare than others, but any individual showing them has a problem. With most human stereotypies, even if they last for a short time the perpetrator is deduced to have some psychological problem (Broom, in press). An occasional bout of stereotypy indicates that a person has a problem at that time; more frequent occurrence of stereotypies is interpreted as evidence of a more substantial problem.

Most stereotypies, even those which involve little movement, such as sham chewing in pigs, or those which are prolonged such as the elaborate movement routines of some caged mink, are easy to recognize if behaviour is observed carefully. These are sometimes ignored by those who keep animals, however, and may be taken to be normal behaviour by those people if they see only disturbed animals. For example, zoo keepers may see route tracing by cats or bears, laboratory staff see twirling around drinkers by rodents, and farmers may see bar biting by stall-housed sows without realizing that these indicate that the welfare of the animals is poor. A greater awareness of the importance of stereotypies as indicators of poor welfare is resulting in changes in animal housing. More complex environments which give the individual more control and hence result in the occurrence of fewer stereotypies are now being provided in good animal accommodation. These environments reduce the incidence of stereotypies by providing animals with more chances of reaching consummation, so that their behaviour does not get stuck in its appetitive phase, as may be the case with some stereotypies. They also give opportunities for a larger proportion of the full behavioural repertoire to be expressed, and for the patterns of movements in the repertoire to be varied. The consequent reduction in frustration and increase in the proportion of an individual's interactions with its environment that are under its control improve its welfare.

6.10.3 Apathy and unresponsiveness

People who are depressed because specific or general conditions of life make it difficult to cope may show reduced activity, apparent unawareness, and lack of interest in the surrounding world. Such apathetic behaviour has also been described in non-human species. The loss of an important companion, whether human or canine, can result in apathetic behaviour in dogs. Animals in inadequate conditions in zoos or farms are also often apathetic. Inactivity and apathy in sows confined for long periods in a small pen or tether stall was described by van Putten (1980) and Wiepkema *et al.* (1983). The degree of apathy of sows in stalls was assessed using measures of responsiveness by Broom (1987) using three different stimuli. Stalled sows were responsive to the arrival of food, but they showed little response to a person who stood in front of them unless that person approached to within 1 m. In a test in which 200 ml of water was tipped onto the back of sows which were lying but awake, those in stalls had a response lasting a median of 27 seconds whilst animals in the same building housed in groups had a response lasting a median of 344 seconds.

Individuals who do not respond to events in their surroundings are clearly behaving in an abnormal and unadaptive way. Unwillingness to explore is often shown by people who are unresponsive to stimuli presented to them. Quantitative measures of responsiveness and explorative curiosity could be more widely used when assessing welfare.

It has been suggested that unresponsive animals may be using endogenous opioids to help them to cope (Broom, 1986b, 1987). A study by Zanella *et al.* (1992) shows that the density of μ receptors in the hypothalamus and striatum of sows is higher in unresponsive sows than in responsive animals. This could be related to self-narcotization in some way, but we do not know exactly what causes these differences in μ receptor density (Section 6.8). Regardless of whether the unresponsiveness is associated with endogenous opioid action, the behaviour of such animals is abnormal and they have substantial problems.

6.11 LACK OF STIMULATION AND OVERSTIMULATION

6.11.1 Lack of stimulation

The effects on behaviour and welfare of enriching the environments of domestic, laboratory and zoo animals are discussed in Chapter 7. Studies of the effects of sensory deprivation on human and other species refer to the same continuum and the measurements used can give information about welfare. Experimental studies on humans are of short duration, but reports from people in such studies and from those who have endured solitary confinement in prison make it clear that it is most unpleasant to have inadequate levels of stimulation. Despite this, and our knowledge of the complexity of life of mammals and birds, many animals are kept in such barren environments that there is bound to be some sensory deprivation, a lack of novelty and few opportunities to explore. As mentioned earlier, various behavioural abnormalities occur when animals are kept in impoverished environments. The consequences of boredom are described by Wemelsfelder (1990) and, whatever the individual feels, we can certainly recognize abnormalities of behaviour in environments which we predict to be boring. Behavioural responses to boredom sometimes seem to be directed towards increasing the level of sensory input, but in the latter stages of stimulus deprivation the most frequent consequences are either stereotypies, inactivity or apathy.

The level of environmental variation below which the effects of stimulus deprivation become evident must vary from one individual to another. Certain people will find too low levels of input which would be quite adequate for other people; there must be similar variation in other species. The differences in responses to barren environments amongst caged animals may in part be differences in coping strategies or in pathological consequences, but some individuals could just be less affected than others. Environmental conditions during early development influence the severity of the effects of sensory deprivation during later life. However, neither early training nor genetic selection can push the individual beyond its biological potential and a profound lack of stimulation is something to which no vertebrate animal is likely to be able to adapt.

6.11.2 Overstimulation

The problems associated with a lifestyle in which there is too much to decide are well known in modern human society. Physical consequences, such as stomach ulceration, have been described in rats and other animals as well as in people. Responses to overstimulation include withdrawal from the confusing part of the environment and concentration on activities with predictable consequences. Here again, stereotypies are a frequent response. This similarity of response to several different kinds of circumstances is further evidence for the concept that it is an animal's lack of control of its environment that is the cause of the behavioural abnormality.

Experimental data on the responses of animals subjected to too much decision making come from studies such as those of Weiss on rats (1971) and from work on 'executive' monkeys. Monkeys trained to follow exacting schedules of lever pressing to obtain resources which were important to them were apparently overloaded in ways similar to the experience of harassed business executives. The monkeys developed physical and behavioural symptoms indicating that their welfare was not good, particularly when the tasks became complex. Even these complex experimental tasks do not mimic the most difficult tasks in life, however, which are arguably those encountered during the establishment and maintenance of social relationships. Social animals must possess elaborate brain mechanisms to deal with these complexities of social life. When problems of social interaction in domestic animals combine with physical and nutritional difficulties, inability to master the situation will often lead to behavioural abnormalities. The pig which lives in a group but shows a stone-chewing stereotypy, and the cow which shows tongue rolling in a cubicle house may be examples of animals trying to cope with overstimulation.

People who are overloaded with decision making are often unwilling to explore or attempt to learn new skills. Reduced exploratory behaviour is also a possible response to overload in other species. Social behaviour in primates, dogs and some farm animals provides examples of having to contend with a disturbingly large array of variables in everyday life. At present, however, we do not know much about the inputs which cause problems or the symptoms of overload in species other than our own. Even in humans we have much to learn about both causation and treatment of such problems.

6.11.3 Problems caused by specific localized stimulation

Localized stimulation usually causes short-term rather than long-term problems for individuals. In some circumstances, however, the stimulation is repeated at frequent intervals over a long period or persists at a low but detectable level. As explained in Section 3.1.2, mild stimulation, if sufficiently protracted, becomes disturbing and noxious. For example, some parasites cause irritation at frequent intervals. The threadworm *Oxyuris* is not painful, but it may cause repeated irritation in the anus which elicits behavioural responses such as rubbing the anus with the limbs or on other objects. Frequent rubbing behaviour indicates continual irritation and this may be associated with more substantial responses.

Horses which show frequent rubbing of the anal region may be responding to *Oxyuris*, to fungal infection of the perineum, to louse infestation or, occasionally, to no obvious cause (Fraser and Broom, 1990).

Sometimes localized, persistent stimulation results from injury. The injury itself may elicit behavioural responses over a long period, even to the point of self-mutilation. In addition, tissue damage can result in the formation of neuromas. These may well cause an increase in behavioural responses ranging from reduced usage of that part of the body to increased aggressive behaviour or stereotypies.

6.12 INTERRELATIONSHIPS AMONG MEASURES

As explained in Chapter 4, individuals vary in the methods which they use to try to cope with difficulties, and in the consequences they suffer due to failure to cope. Welfare should therefore be assessed by a range of measures. This leads to the problems of how to compare the results of different measurements and how to decide whether welfare is worse when, for example, a certain level of adrenal activity is recorded than when a certain frequency of stereotypy or degree of immunosuppression is recorded. This problem is discussed in Chapter 8.

Chapter 7

Preference studies and welfare

If we want to find out what resources and living conditions people need for good welfare, we can study what they choose when given access to alternatives. Once an option is chosen, we must then also take account of the actual effects of having that resource. The assessment of such effects was the subject of Chapter 6, while this chapter is about what is preferred and what is avoided. Observing preferences is also a well-known guide to providing adequately for the animals we keep. Dog owners soon come to recognize the indications given by their dog that it wishes to have food or to go out for a walk. Similarly studies in which farm animals are offered different foods have been of value in deciding which foods to provide and which to avoid. In recent years, sophisticated experimental techniques have been developed which give detailed information, not only about the existence of a preference, but about the strength of the preference.

One obvious factor which influences a person to show a preference for some item or activity is the previous experience of that person. Preferred foods, companions, resting places and so on differ according to early and recent experiences. The importance of previous experience in studies of preference which are relevant to animal welfare is emphasized by Mendl (1990). Dawkins (1977, 1980, 1981) reports that hens which had lived for some time in a battery cage preferred to go into a similar cage rather than to an outside run with grass during the first day such a choice was available to them. This preference was strongly reversed on following days, but the initial effect of the previous experience was clear. Mendl also suggests that animals which have prolonged experience of an environment over which they have little control may not show any meaningful preference if subsequently offered an alternative, because the first environment may have led to apathy and unresponsiveness (Broom, 1986b, 1987) or even learned helplessness (Overmier *et al.*, 1980).

1 TIME AND ENERGY ALLOCATION IN A RICH ENVIRONMENT

When people first encounter a novel, rich environment they usually spend some time exploring. Such behaviour will result in gaining information which will help them to satisfy present or future needs. The possibilities for defending themselves or otherwise responding to danger

short term needs.

may have to be considered; water and food sources must be found; a suitable resting place must be decided upon; the potential for various forms of social interaction have to be assessed. As the environment becomes more familiar there is still an element of exploration, but the individual begins allocating time and energy according to a wider variety of needs. If there are many opportunities for activity, the way in which the person spends time will give an indication of their needs and what they find important. For a person in solitary confinement in a prison cell, on the other hand, many activities could not occur. Information about what conditions are likely to result in good welfare for people is obtained from our knowledge of what people choose to do.

The same approach is relevant to understanding the needs of other species of animals. Part of our knowledge of needs is gained from finding that adverse effects occur if a particular resource, condition, or possibility for action is lacking; part also follows from the observation that animals make efforts to obtain the resource. The idea that dogs need regular exercise stems largely from the observation of the preference of dogs to take exercise. Similarly, most people would consider that birds of most species need to build a nest because they observe that, at times, individuals make considerable efforts to obtain certain materials and use them to construct a nest. The behaviour in the wild of some of the larger animals which we keep in zoos is quite well described, and such information should be used when designing accommodation for captive animals. That is not to say that all of the conditions in the wild, including the possibility of contracting severe disease or being chased and caught by predators, should be reproduced in captivity, but the needs of captive animals can certainly be deduced from studies of wild animals.

Animals kept on farms, in laboratories and as pets should also be observed in rich environments. Species that have been domesticated for thousands of years are different from their ancestors in various ways, so the evidence should be taken from studies of the domesticated strain in varied or semi-natural conditions. It is of particular interest, however, that when detailed studies of feral animals of the domestic strains were carried out, far more similarities to than differences from their wild equivalents were found. McBride *et al.* (1969) found that the behaviour of domestic fowl which had lived unrestrained for many years on an island off the coast of Australia was very similar to that of the Burmese red jungle fowl from which domestic fowl are descended. Similarly, studies of modern pig breeds living in semi-natural conditions in Scotland (Wood-Gush, 1988) or Sweden (Jensen and Wood-Gush, 1984; Jensen, 1986) reveal many similarities to wild boar (Frädrich, 1967).

The fact that the brown rat *Rattus norvegicus* and the house mouse *Mus musculus* are pests has led to the collection of considerable information about their behaviour in the wild. Some early work related principally to how to trap and poison them, e.g. Shorten (1954), but other studies concerned their social behaviour and the ways in which the animals spend their time, e.g. Christian (1961), Calhoun (1962). The descriptions indicate strong avoidance of being in the open; and preferences for staying near cover, going into tubes, exploring, climbing, being with other animals for much of the time and for being able to get away from them at other times. This should demonstrate to any biologist the inadequacy of the average laboratory cage. Our current knowledge of the behaviour of rats, mice, guinea pigs, rabbits and various primate species makes it possible to design much better cages for these animals. Such cages should be tried out, taking into account other factors such as their effects on health and ease of management.

Our domestic animals are undoubtedly different from their ancestors in some ways but, as mentioned above, when they are studied in wild or semi-wild conditions the similarities to their wild relatives are often remarkable. Cats may become feral, and their behaviour in such a state has been described in several detailed studies, such as Natoli (1985), and Kerby and Macdonald (1988). Dogs are less likely to become feral and less evidence is available about them. However there are studies contrasting the behaviour of dogs in rich compared with impoverished environments. Hubrecht, Serpell and Poole (1992) described the behaviour of dogs in animal shelters and laboratories. The most striking findings were the differences in activity levels and in the range of behaviour shown. Solitary dogs were inactive for 72–85% of the time whereas dogs in social environments were inactive for only 54–62% of the time. Dogs kept in groups within a complex physical environment spent more time showing investigative behaviours than did those housed individually in bare kennels with very limited contact with the dog in the next kennel. Similar patterns of behaviour have been reported for a variety of other animals in impoverished environments.

One of the best examples of a study of animals in a rich environment, which has been followed up by use of the information in the design of housing, is that of Stolba and Wood-Gush on pigs. Stolba (1982) and Stolba and Wood-Gush (1989) developed a 'pig park' which consisted of an enclosure with fields, forest and a shrubby area. The enclosure was provisioned with the amount of food that the pigs would have received in commercial housing. Family groups of pigs were observed and their behavioural repertoire was found to consist of 103 separate activities. The allocation of time to the main categories of activity is reported in Table 7.1. Some of these behaviours are infrequent but are apparently important to pigs. It is interesting, for instance, to note that long periods were spent grazing and rooting and far smaller periods were spent lying than is observed in most farm housing. The importance of exploratory and investigative behaviour for farm animals is emphasized further by Wood-Gush and Vestergaard (1989).

Table 7.1 Mean time allocated to different types of activities during daylight by pigs in a rich environment (from Stolba and Wood-Gush, 1989)

Activity	%
Grazing	31
Rooting	21
Locomotion	14
Lying	6
Orienting to stimuli	4
Nosing	4
Agonistic behaviour	4
Sexual behaviour	4
Social behaviour	3
Marking	3
Standing	2
Drinking	<1
Manipulating	<1
Comfort	<1
Suckling	<1
Eliminative behaviour	<1

The careful observation of animals in complex environments also gives specific information about what they choose to do in particular circumstances. For example, in a study of pigs in fields and woodland, Jensen (1989) described the choice of farrowing locations and the nest building behaviour of sows. Observation of 60 farrowings showed that the site chosen for farrowing was normally at some distance from the group resting area, and was in small wooded areas with fields around them more often than in open fields, fir plantations or marshy areas. A substantial nest was usually built, with some time being spent selecting nesting materials.

Clear behavioural preferences of domestic animals are also revealed in almost any investigation which includes a description of how the animals allocate their time and energy when living in a varied environment. When sheep in a grazing area in hot weather have access to shade from trees or other sources, they usually spend some of their time in the shade. In order to go into shade they must expend a certain amount of energy in walking and they will have to forgo feeding for some time. The time spent in the shade varies from individual to individual (Sherwin and Johnson 1987). Thirty-two animals in a flock of 39 spent 70–80% of the hot part of the day in shade, but 3 animals in the group spent only

40–50% of this time in the shade. This variation was partially explained by the group's social structure.

7.2 EXPERIMENTAL STUDIES OF ANIMAL PREFERENCES

Very many investigations of the preferences of animals have been carried out, but those where results can be used to improve to animal welfare have mainly concerned farm animals. The simplest kind of experiment is that in which animals exhibit a preference by carrying out a simple motor activity. Hughes and Black (1973) wished to find out which kind of flooring would be chosen most by hens. They put hens in a cage with three types of flooring: hexagonal wire mesh, coarse but more rigid rectangular mesh and steel sheet perforated with large holes. The hens stood for longest on the hexagonal 'chicken' wire, probably because this gave the best support to their feet. In a similar preference test, piglets were found to lie for longer on perforated plastic or concrete than on wire mesh, and to spend more time in a pen with a container of straw than in pens without this (Marx and Schuster, 1980, 1982, 1984). There are also other simple preference studies, such as those indicating the preference of rats for solid rather than wire grid floors for resting (Manser and Broom, in prep.) those involving selection of one food when offered many, and those involving a choice of one mate rather than another. Tilbrook and Cameron (1989) put a ram with three ewes which had long fleeces and three which had been recently shorn; the ram preferred to mate with the woolly ewes.

In all of these studies an animal exhibited a preference, and had to allocate a small amount of time and energy in order to do so, but it was not clear how important this choice was to the individual. The likely relevance of a preference to the welfare of an animal is indicated in a more convincing way if that individual has to overcome some obstacle, do some work, or take some risk in order to obtain or achieve whatever is preferred. Ways of finding this out experimentally are considered in the next section.

7.2.1 Assessing the importance of preferences

The experimental studies below illustrate techniques requiring progressively greater strengths of preference. In the first studies an individual exhibits a preference which can be reversed by giving the animal an opportunity to obtain a resource or achieve an alternative. When Millam (1987) offered turkey hens a row of nest boxes in which to lay, some hens showed a preference for boxes at one end of the row (42% of eggs laid) over those in the middle (33% of eggs) or at the other end (25% of eggs). The preferred end was then illuminated at 650–1000 lux, after which 24% of eggs were laid there, while the proportion of eggs laid in middle at 50–50 lux was 37%, and the proportion laid at the other end was 39%. Thus the brightest light was avoided and the preference was reversed.

A more sophisticated reversal of preference experiment, which effectively involved balancing one preference against another, was used by van Rooijen (1980, 1981). He observed that pigs, in particular the gilts which he was studying, when offered several individual, freely accessible pens preferred to lie adjacent to a pen occupied by another gilt. He assessed the strengths of the gilts' independent preferences for different types of flooring, and then balanced these against the social preference. An earth floor was preferred to a concrete floor sufficiently strongly to counteract the social preference (Fig. 7.1). However the preference for straw over wood shavings as a bedding material was not strong enough to overcome the preference for being near another gilt (Fig. 7.2). A further example of a study in which different preferences were balanced is that reported by Dawkins (1983) who found that hens preferred a cage with litter to a wire-floored cage but if food was put into the wire-floored cage it was strongly preferred.

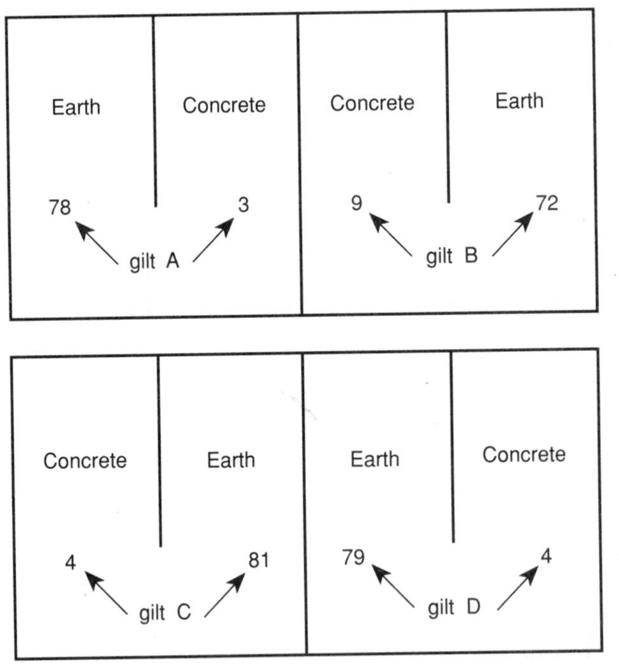

Figure 7.1 Young female pigs (gilts), given the choice of lying on one of two different floors either nearer to the neighbouring gilt or further from her, spent longer (figures quoted for duration in hours) on earth, even at the expense of being further from the other gilt (from van Rooijen, 1980).

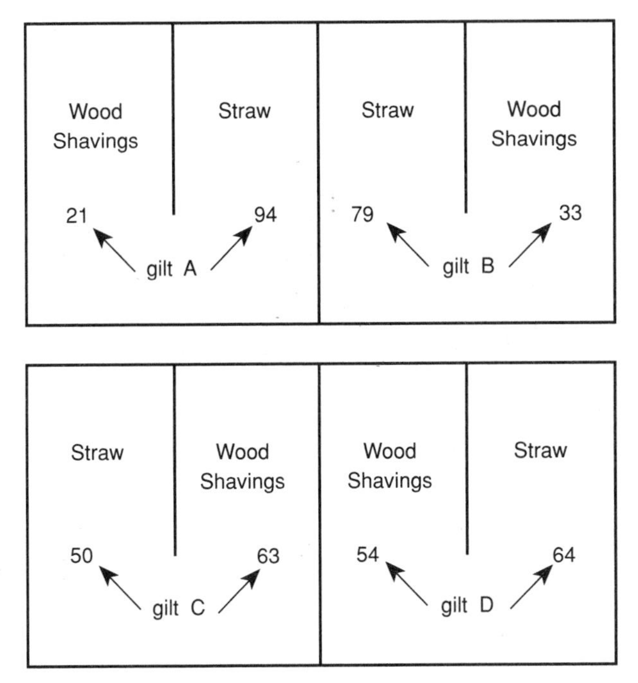

Figure 7.2 Gilts offered the choice of straw or wood shavings as bedding material did not have a strong enough preference for straw to overcome their preference for being near the other gilt (figures quoted for duration in hours) (from van Rooijen, 1980).

In the experiments described so far, the action which the animal had to perform to achieve its objective was merely to go to a place. If that place can only be reached by using extra energy (in a way contrived by the experimenter), the amount of energy which the animal uses to reach its objective can be assessed. This is the basis of a number of studies in which animals can reach an objective only by walking on a floor which moves them away from the objective when they stop walking. Such an apparatus can be disturbing to the animal, but some individuals can adapt to it and then demonstrate their eagerness to reach the objective. A related experimental procedure involves training an animal to push through a door in order to reach an objective and then increasing the force required to open the door. Duncan and Kite (1987) found that hens and cockerels would push open a weighted door in order to gain access to some resource on the other side. Cockerels would push open a heavily weighted door to reach a hen on the other side but hens would not push open a weighted door to reach a cockerel. Petherick and Rutter (1990) designed a door which hens had to push for 13 Newton seconds before it would open and give access to food, and found that hens deprived of food for 43 hours were faster at successfully carrying out the task than hens deprived of food for 12 hours. Clearly these techniques are beginning to give some information about the strengths of animal preferences.

7.2.2 Operant techniques in the assessment of preferences

A more complex way of assessing the importance of resources to animals is to use an operant behaviour which is not directly related to the objective. An operant procedure is one such as that used by many psychologists in which a rat is required to press a lever in order to obtain a food reward. Rats would not normally associate a lever with food but, once the animal associates the action with the arrival of food, its desire to obtain the food can be assessed. The first requirement is that the animal can learn to carry out the operant procedure in order to obtain the reinforcement. The second is to record how often the operant procedure is performed for a particular reward. The pioneer of such studies in relation to understanding the needs of farm animals was Baldwin (e.g. 1972, 1979; Baldwin and Start, 1985). Sheep and pigs learned to operate a switch for food, light or heat (Fig. 7.3) for example by putting their nose in a slot and breaking a light beam monitored by a photocell.

Figure 7.3 This pig is about to push the black panel and thus operate a switch. Pigs learned to press for food, for a period of heat, and to switch lights on or off (from Baldwin, 1979, with the author's permission).

Sheep learned that they could increase their ambient temperature by carrying out this operant procedure and hence it was possible to discover what temperature they preferred. As might be expected, the preferred temperature was higher in a recently shorn sheep than in one which was in full fleece. When young pigs in a group were able to switch heaters on or off in this way, they maintained an ambient temperature at night, when they huddled, that was 11 °C lower than in the daytime when they moved about. Indeed, the temperature chosen at night was substantially lower than that generally recommended for pig housing (Curtis, 1983).

Once it is established that an animal will carry out an operant procedure like pressing a lever or plate in order to obtain a reinforcement, it is possible to withhold the reinforcement until the operant procedure is performed several or many times. This can give a measure of how hard the animal will work for that reinforcer. Many studies with laboratory animals show that they will continue to work for food rewards by pressing a lever even if they have to press it 2, 5, 10, 20 or 50 times. In each experiment the number of times that the lever must be pressed for a constant reward is called the 'fixed ratio'. If the rate of receiving food rewards is plotted against the rate of lever pressing (Fig. 7.4) the plot obtained is called a demand curve (Lea, 1978; Staddon, 1980). If a flat or slowly changing rate of reward is seen (Reinforcer 1), the animal is said to

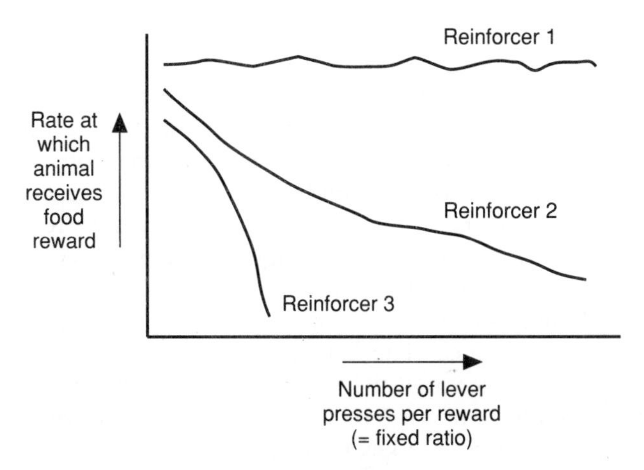

Figure 7.4 Results of experiments in which an animal receives one of three different positive reinforcers when it carries out an operant procedure, such as pressing a lever which it has learned to associate with rewards. When the number of lever presses for each reward is increased (see above), the animal increases the pressing rate and hence maintains the rate of receiving the reward for Reinforcer 1, but does not maintain it for Reinforcer 2, and compensates little for the increased pressing rate necessary for Reinforcer 3.

have an inelastic demand for the resource. If the reinforcer is not sufficient to cause the animal to respond at a high rate, then the demand curve drops as the fixed ratio increases (Reinforcers 2 and 3).

As Dawkins (1983, 1988, 1990) has pointed out, such studies give information about how important a resource is to an animal. This is valuable information in welfare assessment because the fact that an individual is working hard to obtain a resource tells us something about its attempts to cope with its environment and hence its welfare. The actual shape of a demand curve, however, will differ according to the size of the reward, the needs of the animal for the resource and the effort or time required to carry out the operant procedure in relation to other demands upon the animal. The demand curve for increasing fixed ratios of lever pressing for very small amounts of food might be flat if the animal has nothing else that it must do, but steeper if this activity is in competition with a drive to pursue other objectives at the same time.

Another way of using the same kind of apparatus to measure the importance of a resource for an animal is to increase the number of lever presses needed for reinforcement continuously during each session. For example, Lawrence and Illius (1989), working with food-restricted pigs, found that the animals would press a panel for 6 g food pellets at a fixed ratio of 10 presses per reinforcement. When the pigs needed to press the panel once for the first pellet, twice for the second, three times for the third, and so on, they would continue up to 30 presses for a pellet if they had been fed to only 40% of appetite beforehand.

We need to know how important space is to animals when we design housing conditions. As a consequence, preference experiments have been carried out on the space requirements of hens. Hughes (1975) and Dawkins (1976, 1977) found that, once hens were accustomed to having access to a large cage, animals previously kept in a battery cage preferred a large cage to a smaller one. In an attempt to quantify the strength of this preference, Lagadic and Faure (1987) designed an apparatus in which four hens were able to extend their cage from 1600 cm^2 to 6100 cm^2 by repeatedly pecking at a key. The key-peck caused one wall of the cage to move back a short distance. The hens stayed in cages of 1600–1800 cm^2 for 50% of the time, increased the size to over 2500 cm^2 for 25% of the time and to over 4200 cm^2 for 10% of the time. However the hens were not always willing to press the key to move the wall, perhaps because this was a complex situation for the hens, with social factors, a moving wall, fluctuations in activity and requirements for space all playing a part in the result.

The value put on earth for rooting and straw for nest building by pigs has been the subject of several studies. Wood-Gush and Beilharz (1983) found that piglets in small cages spent some time rooting in earth if it was provided and Hutson (1989) found that piglets would repeatedly lift a lever in order to gain access to earth. Six pre-parturient sows given access to a pen with an earth floor all dug an earth nest and farrowed in it (Hutson and Haskell, 1990). If access to the pen with earth in it was dependent upon lifting a lever, using a progressive ratio to assess the desire to get into the pen, only half of the animals made nests and only two farrowed in the pen. This may have been due to some sows being uncertain about whether or not they would always be able to gain access to their nest and piglets.

A further study by Hutson (1992) showed that pre-parturient sows worked much harder for food than for access to straw. Arey (1992), however, found that sows readily pressed a panel on a fixed ratio of 10 for access to an adjacent straw pen in preference to an adjacent empty pen (Fig. 7.5a).

If straw and food were put in different pens adjacent to the starting pen (Fig. 7.5b) two days before farrowing, choices of straw and food were equal with a fixed ratio of one, but food was preferred with fixed ratios of 50–300, depending on the willingness of individuals to press. On the day before farrowing, however, straw appeared to be as important to the animal as food (Table 7.2).

7.3 DO PREFERENCE STUDIES TELL US WHAT IS IMPORTANT FOR ANIMALS?

A fundamental question about the value of preference tests in animal welfare is whether or not animals make choices which benefit them. The actual choosing should result in some immediate improvement in welfare, but this cannot be assured, and it may even be offset by other adverse effects of the choice. Natural selection should result in animals showing behaviour which increases fitness, so some mechanisms promoting efficient selection of resources are likely to exist. This accounts for simple mechanisms like sodium appetite and more complex preferences like those for the sort of hiding places which will reduce predation risk. Strong preferences for opportunities to display maternal characteristics, have social companions, be

Table 7.2 Pre-parturient sows: presses on panel to gain access to straw pen or food pen (from Arey, 1992)

Presses per reinforcer (fixed ratio)	*Day before farrowing*	*Presses*	
		for straw	*for food*
1	-2	17	21
50–300	-2	2.6	11.4
50–300	-1	16.4	17

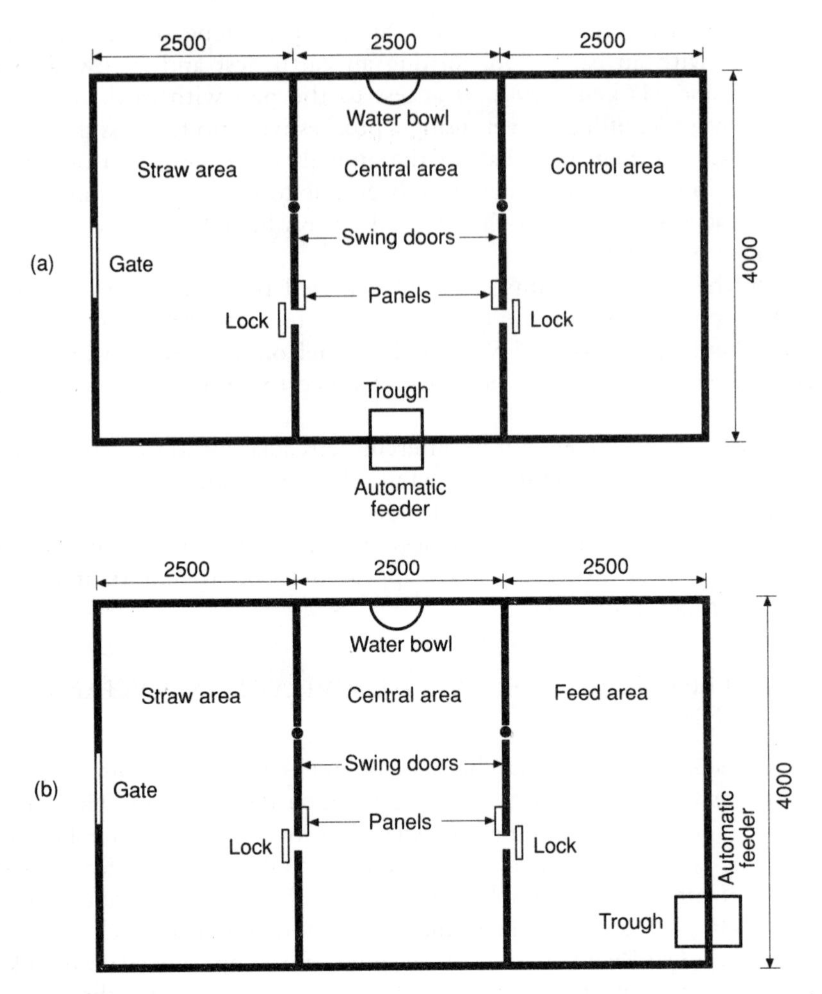

Figure 7.5 By pressing the panel on the door, pre-parturient sows put into the central area could gain access to (a) the straw area or empty control area, or (b) the straw area or feed area (from Arey, 1992).

able to explore the immediate environment, and have particular conditions in which to lay eggs or give birth are also examples of animals' basic biological characteristics which increase fitness. These preferences can be sufficiently strong to become established as 'biological needs' which, as discussed in Chapter 4, can be satisfied only by the animals being able to perform specific behaviours. This argument for animals showing preferences that will be beneficial is supported by a wide range of observations of preferred activities which apparently help the individual or promote the spread of genes carried by that individual within the population.

Despite the general rule that animals prefer what is good for them, Duncan (1978) points out there are examples of preferred resources or activities causing harm. Grazing animals sometimes select poisonous plants and some rats offered a wide range of different kinds of food may choose solely chocolate bars which do not provide adequate nutrition (Rozin, 1976; Broom, 1981). Rats will even tolerate extreme cold to eat such food, despite a balanced diet being available in the warmth (Cabanac and Johnson, 1983). Humans may eat so little that they die or so much that their health is damaged, and they may have strong preferences for drugs which eventually kill them. In some of these examples there is a short-term reward, so the showing of such a preference may be beneficial in certain situations, but have a long-term adverse effect. Early experience can have a strong effect on preferences and could result in a less than optimal resource or activity being preferred. As is evident from these possibilities, not every preference is beneficial.

Dawkins (1988, 1990), in a discussion of such problems, concludes that the solution is to take into account more than one measure of welfare. Preference studies should be combined with the use of indicators of poor welfare. Some difficulties in the interpretation of preference studies result from problems in experimental design, as mentioned elsewhere in this chapter, and other difficulties are discussed in detail by Dawkins. With care, most of these problems are soluble, and there is no doubt that preference tests are an important tool in the assessment of welfare. They have particular value in the planning of better housing facilities for animals, and in the selection of management procedures. Once designs for such facilities have been developed, the new system can be compared with alternative systems using both preference tests and other measures of welfare.

Chapter 8
Ethical problems concerning welfare

In the last three chapters, tests have been described which can be used to assess the effects of unfavourable conditions on an animal. Unfortunately these alone do not solve problems concerning welfare. In the first place, it is still necessary to select the most appropriate tests from amongst many. Indeed, new tests may need to be developed. But at least it is now conceptually possible to understand, and technically possible to quantify, some of the strains on an animal due to particular factors in its environment.

However, another major problem about the application of such welfare tests has not been addressed. Furthermore, this question is at least as important as the assessment of stress and strain. The question is, how does one decide when and whether imposing a particular strain on an animal is justified? How poor must the welfare be before the use of such a treatment or condition for the animal is considered unacceptable? As emphasized earlier, it is essential that the procedure of welfare assessment and the ethical decision are separate; but how is the ethical decision reached?

To make this point clear, consider two examples. Slaughtering an animal, if done to alleviate human starvation is, in the minds of many people, justified. Slaughtering an animal to satisfy a person's penchant solely for killing is, for many people, not justified. The acts may be identical, the objectively assessed strain on the animals may be identical, but the justification for the acts is different. The law in most countries treats the two situations in very different ways. There are detailed regulations aimed at minimizing suffering in the case of killing in a slaughterhouse but often no regulations limiting the suffering of animals which are hunted. Indeed the hunted animal can not only be killed inhumanely but it can be subjected to extreme fear, pain and other suffering without any legal protection. The laws are based on quite different assumptions.

If a decision on this issue seems straightforward, consider a dilemma any one of us may have to face. You, or someone in your family, has a painful and debilitating medical condition which you are told may be curable by a radical new therapy. You are also told, however, that it is possible that the treatment could be more distressing than the complaint, so medical advisors suggest it should first be tested on laboratory animals.

Would you feel justified in imposing this test for your own benefit if it caused slight effects on an animal's welfare, or quite severe effects, or very severe effects? Would it be reasonable to cause such effects in a single mouse or 100 mice, 1 or 100 dogs, or 1 or 100 chimpanzees? To be realistic, one probably needs to have in mind a particular person such as a daughter or father to test the true strength of a commitment to such a decision.

Agonizing dilemmas arise from time to time where one must decide between the welfare of humans and animals, or between one animal species and another. For example, situations arise in which pest insects might be exterminated because they are eating plants used as human food; in which experiments on a few animals might improve the welfare of many more animals; or in which pet animals should perhaps be killed because they are preying on other species. All these problems present us with dilemmas about animal use and control for which there can be no simple solution. The best we can hope for is that we can identify some realistic options for solving the problem, understand the impositions on the animals concerned, and implement the better rather than the worse solutions.

Clearly, we have moved into the realm of value judgements about animals and their welfare. This chapter discusses some of the ways that these crucial problems can be approached.

8.1 VALUE SYSTEMS

Decisions about what might be considered justifiable impositions on animals depend on our value systems for animals. In simple terms, these systems operate at two levels. First, each person has a private attitude, which they may ascribe to belief, intuition, learning or experience. For many people, this aspect of their personal philosophy is probably not clearly defined, or easy to enunciate; such people simply 'like cats', or 'enjoy hunting'. An increasing number of other people have taken up one of the codified philosophies about animals which is much more clearly structured, and may have further developed it by adding ideas from reading, observation and discussion.

Society also comprises various communal value systems. These may arise from family, religion, culture, or some other social belief. They will rarely be upheld unanimously in the population, but may be supported by a sizeable proportion of it and perhaps be reinforced by legislative control. For example, some societies demand a certain minimal level of treatment for animals, and have legal injunctions against cruelty. On the other hand, some societies accept and perhaps expect that people will impose on animals as part of rituals such as bull fighting, horse racing, or particular procedures of animal slaughter or sacrifice.

Judgement as to whether acts causing distress to animals are justifiable or not is made by relating them to both one's personal philosophy and to the current expectations of society. The objective measures of animal stress and welfare described in earlier chapters can provide part of the answer as to whether society should sanction a particular activity. The other part depends on public and private attitudes as to what is acceptable.

In order to begin an ethical discussion about stress and welfare, we must start with a brief examination of the basic relations between humans and other animals.

8.2 HOW HUMANS IMPOSE ON OTHER ANIMALS – AND VICE VERSA

From a biological perspective, humans appear to behave in the world little differently from any other animal species. Some species including humans occupy dry-land environments but not aquatic ones; others such as fish, planktonic crustacea and whales do the reverse. Humans use certain animal species for transport; so do various parasites, such as some insects and sucker fish. Some humans live on the flesh of animals; some animals live on the flesh of humans. Some humans live in highly structured societies with division of labour; so do some insects, birds and other mammals. We nurture and defend our offspring until they are self-sufficient; so do many other species. Some humans hunt and slaughter more animals than is necessary to provide food; so, sometimes, do foxes and dogs. Animals of many species, including humans, behave as if they experience pain, anger, fear and love. So is there any important difference between man and other animals?

If there were no difference, we might, arguably, not be obliged to treat non-human species in any different way from the way they treat us. We could, with some justification, approach the world as an arena where each individual behaves to enhance the survival of its own genes, or rather where each gene acts in a way which maximizes the chances that it will spread in the population. We have until recently failed to perceive the probable long-term consequences of such actions; the result of a freely competitive approach is becoming increasingly evident in the world at present. Our human powers are so pervasive and crudely controlled that, through unrestrained exploitation, we face a risk of exterminating not only many animal species from which we derive companionship, pleasure and inspiration, but also those that are biologically essential for the future success of our own species.

Some people believe that there is a difference between humans and other animals that is determined supernaturally. For them, the special roles of human and non-human animals are revealed by God.

In a manner that is consistent with both the above proposals, it could be that there is a difference between ourselves and other animals because humans have special abilities that give them advantages over, and therefore perhaps responsibilities for, other species. In this sense, we may be more advanced than other animals and should therefore exercise a stewardship over them with appropriate concern for their lesser capabilities.

Before proceeding, it is worth digressing briefly to acknowledge that it may be presumptuous of humans to consider themselves superior to other species. Why should we feel superior to birds that can fly, to spiders that can hang on ceilings and build beautiful webs, or to the many species that live in environments without despoiling them? But we must leave aside these digressions without deciding whether they are trivial or profound, and modestly seek to identify capabilities which humans possess that are not evident in other species.

At the simplest, there are two: greater technical skills and a larger capacity for complex forward planning. Even these are only matters of degree. Termite colonies display obvious technical skills in constructing nests with temperature and humidity control. Many animals plan ahead when they prepare themselves for predicted danger, eat enough food to last them until the next predicted meal or store provisions for the next season. However, at least from the human perspective, there is a difference between humans and other animals in the complexity, the range, and particularly the combination of skills used by human animals. Only the staunchest animalist could deny our extensive and special, though not necessarily beneficial, influence in the world.

Although the biological world is apparently one in which each species is seeking to perpetuate its kind, the competition between human and non-human animals is unequal. *Homo sapiens* is not necessarily dominant by right, but rather because we have greater technical, organizational, and exploitative skills. These currently appear to be leading us on a biological misadventure toward a rather bleak future. But fortunately humans also have two related characteristics that may stop the drift. Firstly, we may have detected just in time the imbalance between our own and other species and made a turn to biological conservation and recognition of the value of other species. Secondly, it may be that we can use some of our organizational capacity to temper our technical activities with compassion.

No philosophy or legislation could completely prevent humans from imposing on other animals; imposition is a feature of the biosphere. Even vegetarian food is produced at the cost of killing soil fauna, cotton cloth requires control of insect pests, and some species must suffer because there are obligate carnivores. But our use of animals need not be profligate and cruel, it can be economical and caring. We can afford, and have the knowledge, to reduce our impositions on the animals with which we share the world.

Thus the argument turns full circle. For to assess welfare, the plethora of tests described in earlier chapters is available. With them, and only with them, can we have a reliable guide as to how an animal is coping with its environment.

8.3 SETTING LIMITS TO ASSESSED WELFARE

When some human imposition on animals is unavoidable, yet should be minimized, the problem of setting limits reappears. These limits could, theoretically, be absolute or relative. Absolute values of coping ability are difficult, if not impossible, to prescribe, since the responses of different animal species to various impositions are highly variable. In practice, limits are therefore relative, and to some extent arbitrary, though ideally they will be logical and justifiable.

Three comparisons for establishing standard limits for welfare should be considered.

8.3.1 Animals in a natural environment

Impositions on animals could be deemed acceptable if they impose no greater strain on the animals than that which they experience in their natural state. This is an attractive proposition in concept, but, for most species, unrealistic in practice for several reasons. Firstly the strains on free-living animals are not adequately documented. Secondly, intervening in the life of an animal to estimate its level of strain could increase that strain to an extent that would be hard to determine.

A third problem is that livestock and companion animals must have changed during the millennia of domestication. What animal should we compare with our present-day animals? Should a dairy cow be compared with an aurochs or a feral cow, and under what conditions? A free-living dog, with skills honed in the wild to hunt and escape, seems hardly the appropriate animal to indicate, by comparison, whether a house-dog is being provided with appropriate conditions. A final, particularly important problem arises from the fact that many animals die in the wild as a result of disease or predation. Most people consider that whenever we use animals we have obligations to care for them so that they do not die of disease and predation. The idea of relating animal welfare to experience in the wild is probably acceptable only in order to set a minimum standard.

The practical difficulties of setting up welfare standards are formidable. However, valuable information does come from studying untamed animals or feral animals that have returned to the wild. For instance, the observations by Stolba and Wood-Gush (1989) and Jensen (1989) of domestic pigs roaming free outdoors provide valuable indicators of what aspects of an animal's life are of importance to it.

.3.2 Humans under the same strain

A feature of the general indices of stress, strain and welfare described in Chapters 5, 6 and 7 is their applicability to many species, including humans. Levels of plasma cortisol, ß-endorphin, adrenaline, or heart rate are likely to be altered in most animals subject to short-term stress. This parallel response could permit some comparison between humans and other animals.

A human appreciation of what an animal might experience could perhaps be derived by noting which impositions elicit comparable strain. For example, suppose that the behavioural and hormonal disturbance of a calf during dehorning elicited the same response as that measured in a person having an intramuscular injection; in that case, it might be deemed tolerable. If, however, dehorning was comparable to a tooth extraction without anaesthetic, it might be considered unacceptable. However, once again complications intrude for, in a country where people do have tooth extractions without anaesthetic, the judgement as to what is an acceptable imposition on animals would be different. Does the level of acceptable animal welfare depend on the level of human tolerance of discomfort or pain? That would be the logical conclusion if we used this approach.

Comparing species will also undoubtedly highlight disparities between indices, so comparisons will not necessarily agree for all indices. That aside, the approach has the attraction that the impact of a stress on some non-human species can be interpreted in terms that are meaningful to a human.

.3.3 The informed and compassionate arbiter

Perhaps the most realistic procedure for determining limits of welfare is though the judgement of one, or a panel of, human arbiters who take account of the biological characteristics of the animals, including their response to difficult conditions. Such people can weigh up the complexity of factors: the comparison of aggregate discomfort with perceived benefit. Given the biological inclination for animals, not least humans, to pursue their self-interest, the procedure is obviously sensitive to the influences of human inadequacies and biases. Systems to monitor this by giving agreed weighting to various aspects of a decision have been proposed in relation to pain by Bateson (1991). The earlier proposals to use the behaviour of undomesticated animals or human strain equivalents for comparison could also provide independent checks of, and complements to, ethical decisions.

One implication of a scheme using arbiters is that standards of acceptable animal welfare are not absolute, but change with improvements in knowledge of welfare indicators and with society's expectations. As already noted concerning comparisons of strains in humans and non-humans, considerable differences in attitude are bound to exist. This will surprise no one in a world where commendable, agreed international declarations about human welfare are often swept aside by tides of political expediency. Despite dreams that it might be otherwise, acceptable impositions on animals will alter from peacetime to wartime, and in times of plague and famine.

This does not imply that objective analyses of stress and welfare are worth nothing – quite the reverse. For measurements of strain will not change, and evaluation of them will be refined by increases in knowledge: it will be possible to say with assurance whether or not an animal is suffering. Whether that suffering will or will not concern human society will depend on that society's current preoccupations and priorities.

Because of the need to reflect society's views, making decisions about animal welfare must involve an opportunity for society to be represented and cannot rest solely with the scientists making the measurements. In some areas, this mechanism for public representation already exists in the form of membership of appropriate committees (e.g. the Animal Experimentation Ethics Committees in research institutes in Australia). In other circumstances, inspectors answerable to the government, and only more distantly to the electorate, judge what is currently acceptable to society, as is the case with the Home Office Inspectorate in the UK. In all such situations, decisions about animal welfare should be made by people who have a knowledge of – and an empathy with – animals, who know how particular conditions might strain an animal (or if they don't know, could measure it), and finally who are aware of society's expectations, both when these are embodied in laws and when they are otherwise advocated by the community.

8.4 UNRESOLVED DIFFICULTIES

The value of scientific indices of stress, strain and welfare can easily be overrated. Offering a solution to a problem is not the same as solving it in practice, particularly with social issues. Problems of assessment of stress and welfare and judgement about where to draw the line about what is acceptable will continually arise.

Two major problems which can already be foreseen relate to trading off immediate strains for future benefits (see Section 2.5.4). The simpler problem is quite easy to understand in human terms. One consequence of the human capacity to anticipate and plan is that present costs can be weighed up against likely future benefits when deciding whether to accept a particular treatment. Thus, most people are willing to tolerate a vaccination for cholera for protection against a disease that would presumably be much more unpleasant. However, they might be unwilling to accept a replacement of their hip joint, if they believed that the operation had a low success rate.

Can we expect animals to do the same? We have little communication with animals to reassure them that we believe the present imposition is in their long-term interest, apart from verbal and physical encouragement. Making such a decision on an animal's behalf must rest with a human arbiter. What must be hoped is that the human custodian has been assured that the treatment is necessary, that there are no other less stressful alternatives, and that the treatment is imposed with minimal unavoidable suffering. These topics will be discussed again in Chapter 9.

A trade-off that is harder to evaluate in our domestic animals is that which parallels the acceptance by humans of present suffering for future 'satisfaction'. People train athletically and race themselves to physical exhaustion, they tolerate malnourishment so that their offspring might eat, they fight to defend their territory or country. If other species are like humans, do they also accept short-term difficulties as part of the cost of success? Examples from animals in nature indicate that they will fight to defend their territory, their offspring or their mate. But do horses and dogs also derive satisfaction from racing? Do guide dogs, security dogs, and dogs detecting drugs have fulfilment when carrying out their duties, adequate to provide them with an effective trade-off against other activities which they are denied? Or are they effectively enslaved?

There can be no straightforward solution to these and many other dilemmas. A leap of judgement is needed by humans on behalf of other animals. In practice the only safeguard that can be instituted to protect animals against arbitrary imposition of stress is to have such judgements made by informed and compassionate arbiters.

Chapter 9
Solutions and conclusions

9.1 PURPOSES OF STUDYING STRESS AND WELFARE

The present upsurge of interest in animal welfare (Harrison, 1964; Singer, 1990; Serpell, 1986) emerged from public concern about animal care and management. In the public discussion that has ensued, both ethical and biological issues have been raised, leading to protracted and, at times, intemperate debate. The matter is now in need of new analyses that could resolve some of the disagreements and improve conditions for animals.

In this concluding chapter it is appropriate to begin by highlighting two essential links between the biological and ethical problems concerning animal welfare. First, there is the need to accept that although the welfare of an animal is assessed using a knowledge of biology, decisions about what an animal should be expected to tolerate must be based in part on ethics, as discussed in Chapter 8. Second, there is a requirement that any ethical proposals must be biologically sound (Lockwood, 1987).

The requirement for the biological validity of ethical proposals has two important facets. One is that we need to accept that animals always impose on each other. Since interaction involves stimulation, and often constitutes an imposition, animals have evolved to cope with a degree of environmental disturbance. Some stimulation which is disadvantageous in the short term is inevitable, and some of this stimulation may have beneficial effects in the long term. Animals exposed to an assortment of stimuli cope better when they have previously faced occasional bouts of at least mildly noxious stimulation. Conversely, it appears that animals in sensory isolation voluntarily seek stimulation, even when it sometimes seems certain to be unpleasant for them. As a consequence of this, some experiences which might at first be predicted to be mildly stressful are later found to be beneficial.

A second biological fact with ethical implications concerns that most central and influential of stimuli, pain, which can be shown at times to be useful, and should not necessarily be entirely suppressed. Pain is a stimulus that on occasions will contribute to an animal's chances of survival and may improve its welfare in the long run. Pain perception can be important for animals recovering from physical trauma, and perhaps from emotional disturbance too. We need to be aware of the value of the pain system for animals, even though we strive to avoid causing pain on most occasions.

Because exposure to taxing and unpleasant stimuli, perhaps resulting in strain, is apparently unavoidable in any natural environment, biologists should accept that they have responsibilities to understand such matters better. They should seek ways of monitoring the effects and consequences on animals so as to make possible the provision of guidance for those who manage animals and for legislators. It is with these aims in mind that the basic and applied biology of stress and welfare have been explored in this book. We now turn to the last of these aims, that of determining practical applications of monitoring stress and welfare.

2 PRACTICAL APPROACHES TO ASSESSING STRESS AND WELFARE

Despite the fact that people have had some awareness of the nature of stress and welfare for several generations, we have lacked an agreed approach to its formal assessment. Biologists, veterinarians, farmers, pet owners, government inspectors and others involved with animals may be helped by an introduction to the problems of stress and welfare as they are now perceived, and by guidelines to assist them in making systematic and meaningful judgements. Undoubtedly, any approach, including the one outlined here, will need to be constantly reviewed. Aspects of it will become out of date and improvements will certainly be possible in the future.

An ability to assess welfare opens up wider possibilities than might first appear. With a capability to monitor stress and welfare come opportunities both to avoid some stressful situations and to reduce unavoidable causes of poor welfare. Investigation of a welfare problem should involve three aspects: an initial check of whether poor welfare can be prevented, attempts to reduce effects that cannot be avoided and, finally, some monitoring of the strain and coping responses induced by any remaining stress and other effects so that a practice which induces unacceptably poor welfare can be terminated.

Though interrelated, these three types of procedure will be discussed in turn. It will become obvious, and is not surprising, that the practices advocated are variants of common-sense protocols that are already used where welfare is a problem.

2.1 Avoidance of stress

In theory one could compile lists of relevant stresses for every genus of animal under human management. For each species, breed and individual there could be further extensions of those lists. But itemizing stresses even for the commonly encountered animals would require a book far longer than the present one, and be quite impracticable.

However, for some categories of animals, certain stresses have been identified. This is particularly so with farm livestock and laboratory animals, for which government offices or independent organizations have drawn up lists of requirements in the form of welfare codes.

Recommendations for the welfare of farm animals in various developed countries are based on those of the Standing Committee of the Council of Europe Convention on the Welfare of Animals Kept for Farming Purposes. In some guidelines, such as the welfare codes used in some countries, a general approach to specifying how to avoid stress is also used. The UK Ministry of Agriculture, Fisheries and Food Welfare Codes list five basic freedoms which should be given to animals. The version of these detailed by the Farm Animal Welfare Council in 1992 is listed below:

1. freedom from hunger and thirst – by ready access to fresh water and a diet to maintain full health and vigour;
2. freedom from discomfort – by providing an appropriate environment including shelter and a comfortable resting area;
3. freedom from pain, injury or disease – by prevention or rapid diagnosis and treatment;
4. freedom to express normal behaviour – by providing sufficient space, proper facilities and company of the animal's own kind; and
5. freedom from fear and distress – by ensuring conditions and treatment which avoid mental suffering.

These freedoms are described as being ideals which anyone with responsibility for animals should aim to provide, and it is further explained that animals will be better furnished if those who have care of livestock practise:

1. caring and responsible planning and management;
2. skilled, knowledgeable and conscientious stockmanship;
3. appropriate environmental design;
4. considerate handling and transport; and
5. humane slaughter.

These lists identify the principal requirements of animals in relation to significant environmental factors to which they have to adapt, and the obligations of people towards the animals.

An observer of society might cynically point out that we do not yet manage to provide such freedoms for all human populations in both 'developed' and 'undeveloped' countries, but this does not alter the desirability of trying to do so for our farm animals.

Management systems for animals evolve with time, and sometimes include practices for which justification has now changed, or perhaps never actually existed. A management practice under scrutiny for its welfare implications should be critically assessed to ascertain that the procedures are really necessary. For example, it has been the custom to castrate many farm animals and to dock the tails of certain breeds of dogs. Some such practices serve no useful purpose, or have been reassessed – balancing the effect on the animal against the extent of the advantage to man – and are now being abandoned.

People involved with animal management should also keep abreast of technical developments which can circumvent welfare problems. In the past, science and technology have rightly been accused of being part of the cause of welfare problems. However, they can also help to solve them. Welfare problems can disappear as a result of technical innovation. For example, trends towards close restraint of pigs to control their food intake are being reversed now that feeding can be adequately controlled in large pens by computer systems signalled from electronic collars and by the use of group housing with individual feeding. Modern telecommunication systems are also permitting sales of livestock to be carried out on the farm without the need to put animals in sale yards, thereby avoiding some of the undesirable impositions on farm animals during transport.

.2.2 Reduction of stress

It has been argued earlier that some impositions on animals are unavoidable and that some degree of stimulation, perhaps occasionally considerable, of animals is not only normal but beneficial. So, after the first move in assessing an animal welfare problem of removing avoidable stresses, a second move is to bring environmental stimulation into the range that is less than the intensity that constitutes stress, yet above the critical lower level that constitutes sensory deprivation.

Two procedures are involved in adjusting the level of stimulation in this way. Some evaluation must be made of the success an animal is having in adapting to its environment; that is, how much it is having to do to cope, and the extent to which failure to cope is occurring or is likely. As explained earlier, a number of tests may be required to indicate this with confidence. Where the species and the situation are well understood, such as with laboratory animals, and in some farm practices and zoo environments, it may be possible to evaluate the causative influences instead of the consequences for the animal. Thus instead of monitoring plasma cortisol levels, it will often be more realistic to check that the animal has appropriate nutritional, physical and social conditions, most obviously by checking the appropriate welfare code.

The second judgement required is how long a particular level of coping activity can be tolerated and whether it will lead to a reduction in biological fitness. This judgement is more difficult to make confidently as there is far less published information for guidance. Naturally, where a source of poor welfare can be identified, it should be removed if practicable.

9.2.3 Monitoring welfare

Monitoring stress is often an integral part of avoiding and minimizing it. Stress measurements, which are indicators of welfare, are essential for obtaining the information needed for framing welfare legislation. Attitudes to animal welfare cannot be set by law, but procedures to prevent poor welfare can. The complex biology, terminology and ethics have been discussed earlier in the hope of clarifying the situation for legislators. These collectively should provide some understanding of the objectives, methods and possibilities of animal welfare science, and simplify the task of representing accurately what the electorate wishes to have as its public code of animal welfare.

The concepts underlying monitoring of stress and strain may be relatively simple, but the practice is not. Each group of animals, and in some cases each animal, has a unique pattern of response to its environment. There are widely applicable indicators of poor welfare, but variations among groups of animals and situations lead to differences in the measures which are most relevant. In the absence of a large range of data, a general approach based on sound biology and demonstrated practicality must be used.

Schemes based on single values may be useful but, if possible, an aggregate of several measures is preferable. The values can be of potential stresses in an environment, that is, factors that can be measured without the animals being present. These may be known from previous studies which have shown what factors lead to poor welfare, including the various disturbances detailed in Chapters 5 and 6, and to avoidance behaviour. In the same way, earlier studies of animal preferences may have given indications about what factors should be good for the animal. Alternatively, and somewhat more reliably, one could determine directly the welfare of particular animals in the environment under study. It may even be possible to combine these measures into a stress or welfare index.

To see how these approaches work in practice, we could consider examples as different as the intensive farming of pigs and the introduction of an exotic species, say a lizard, into a zoo. Because the analyses are based on common biological principles, the two approaches would be virtually identical.

Even before animals are present, consideration should be given to the stresses to which the animals will be exposed. The people responsible for their husbandry should consider all aspects of the environment in which the animals will be kept, particularly the physical, nutritional and social factors. Information may be available in welfare codes and the Universities Federation for Animal Welfare (UFAW) handbook for farm animals, zoo animals and laboratory animals. Where no such guidance is available, extrapolation will have to be made on the basis of requirements of related species, plus knowledge of the animal's natural habitat. These extrapolations should be checked by monitoring the animals once they are introduced.

The principal biological requirements can be determined before the animals arrive. They might be structured along the following lines:

1. **Physical characteristics;**
   (a) quantity (x m^2/animal);
   (b) quality (protection, drainage, structure, surfaces required, specific objects required);
2. **Nutrition**;
   (a) quantity (x g/day); frequency of feeding;
   (b) quality (roughage, similarity to natural food for zoo animals);
3. **Thermal range**; maintained within x – y °C, shelter;
4. **Social interactions**;
   (a) opportunities for grouping, grooming, etc;
   (b) optional protection from conspecifics or predators

Some of these requirements may be regarded as absolute, others may be given graded importance. In each category, a narrow range may be regarded as ideal, a wider range as tolerable. Outside the wider range, conditions must be deemed unsuitable. For instance, space requirements become progressively less suitable as they decrease below a nominated level. Food quality optima should be decided by taking account of that available to the animal in its ideal natural conditions, and of the animal's digestive functioning and behavioural adaptations to forage. An artificial diet may pose some problems for the animal, so consideration should be given to offering the animal food in the manner and at the rate at which it is available in its natural state.

Once the environment itself has been evaluated, the animals can be introduced, and the conditions of the animals monitored. The factors surveyed should reveal the animal's responses to any factor that might remain as a stress. Items on a list of causes of poor welfare could be given a weighting to reflect their presumed importance. Indices of poor welfare that would be appropriate include growth rate, incidence of disease, reproductive rate and incidence of abnormal behaviour. Incorporating physiological measurements that might indicate strain, such as heart rate, respiratory rate, plasma glucocorticoid concentration, and so on could add precision, but would be justified only if the animals were not disturbed by the monitoring procedure.

These procedures may seem little more than guidelines for elementary animal management. So is it necessary to draw up a formal procedure for monitoring standards of animal care? Regrettably it is, because people's desire to keep animals often outstrips their knowledge of the animal's biology, and sometimes their appreciation of an animal's suffering. Guidelines, whether they are laws or codes of practice, establish standards for the requirements for humane animal care. They codify what is commonly left to human judgement. In so doing they can draw attention to those instances where zoo operators, farmers, pet owners or other animal owners let standards slip. Humans have difficulties in setting benchmarks for their own animal husbandry. This is sometimes, but by no means always, because of economic pressures. Cases are periodically publicized of people keeping large numbers of cats, dogs, or horses out of 'affection' for them, despite the fact that the animals are neglected and debilitated.

A further benefit of formalizing standards of welfare is that this gives us the ability to compare norms of management at different venues. Indices of welfare are important when monitoring or trying to improve the quality of life of the animals which we keep. If goals have been identified and plans of development agreed, systematic studies of welfare can be carried out and new housing or management methods progressively introduced so that there is a steady improvement in animal welfare.

In the establishment of animal welfare codes for various species there are opportunities for valuable cooperation between specialists and those people, such as members of canine associations, sheep farmers and so on, who have special affinities or involvement with a particular type of animal. These groups should seek to establish agreed codes of practice for these animals. For example, most sheep farmers know how frequently their animals should be inspected to avoid problems, and the consequences of different sizes of spacing both indoors and on vehicles when the animals are transported. In conjunction with specialists who have knowledge of experimental studies of various procedures and systems, practicable and humane methods can be agreed. Similarly, pigeon fanciers might confer with veterinarians and other biologists on how often their birds should be raced, what housing the birds should have, and when they should be retired. Wildlife rangers, physiologists and nutritionists could suggest how wild animal populations should be controlled and when food supplements should be given. Through such collaboration, the widest sources of knowledge could be tapped, and the greatest chance given for the implementation of mutually agreed practices. Ideally, the standards set by biologists and animal owners will become progressively better substantiated, as they are in turn reviewed, tested and improved. By such progression, a welfare code could be established or refined and, if necessary, legislative control could be implemented.

9.3 CONCLUSIONS

It has been timely to reconsider stress because of continued calls to quantify animal welfare to provide the basis for 'sober professional judgements based on scientific data as a counter to emotionally based anthropomorphic judgements' (Rushen, 1986c). Stress is a term employed widely, if loosely, in both human and animal contexts, but it has been vigorously criticized, to the point where some have advocated that it should not be used at all (Rushen, 1986c; Charlton, 1991). As explained in earlier chapters, much of the criticism of the usage of the term is justified. The options for the future of the stress concept and its consequences are as follows.

1. **Discard the word 'stress'**

This is unworldly. The word will not disappear. If it did, another equally problematic word would be coined to replace it.

2. **Put 'stress' in the category of words that are unquantifiable**

Some collective words describe real and powerful forces that influence the lives of both humans and other animals, and are useful, though cannot be evaluated. Perhaps stress should be included with collective words, such as comfort, unhappiness or love. This is an enticing escape route but, for scientific purposes, unacceptable for the reasons given below.

3. **Define the term accurately and try to unravel its biology**

Despite the formidable problems to be faced, this seems the only realistic solution as long as there remain outstanding problems concerning the welfare of man and other species. The final compelling reason for persisting is the need for some degree of quantification upon which to base legislation. Means of quantifying stress and welfare have been developed and need to be developed further. Some of these means are specific to particular species and situations, and the construction of indices may be complex. In addition, new methodology will appear. Hence, the interpretation of stress and welfare measurements will be disputed. Nonetheless, the alternative is to make only vague, unquantified statements about the state of an animal's mind and body. It would then be impossible to avoid making 'emotionally based anthropomorphic judgements', which we began by trying to avoid.

This book has outlined not only the moulding of theoretical ideas from biology and ethics into practical applications, it has also presented a model of animal welfare which we hope is sufficiently robust to provide a base to which future research data can be added and is sufficiently clear to be used as a basis for necessary legislation. The details must continue to be tested and refined, so that the various aspects can be accepted, rejected or improved as necessary.

The stage has now been reached where people whose lives intermingle with those of non-human animals must accept three challenges:

1. to acknowledge that social evolution has changed human relations with animals, often to the detriment of animals, and that a review of the situation is timely;
2. to learn how science has uncovered explanations of some of the responses of animals to the various problems in their lives, the critical values that mark the limits of those responses and the consequences of exceeding these limits; and
3. to refine the measures of stress, strain and welfare of animals so that they can be used to improve the relationships between humans and other animals to a level that is appropriate in an informed and compassionate society.

Glossary

Abnormal behaviour, Aberrant behaviour Behaviour which differs in pattern, frequency or context from that which is shown by most members of the species in conditions which allow a full range of behaviour.

Adaptation (1) At the cell and organ level, the waning of a physiological response to a particular condition, including the decline over time in the rate of firing of a nerve cell. (2) At the individual level, the use of regulatory systems, with their behavioural and physiological components, to help an individual to cope with its environmental conditions. (3) In evolutionary biology, any structure, physiological process or behavioural feature that makes an organism better able to survive and to reproduce than other members of the same species. Also, the evolutionary process leading to the formation of such a trait.

Aggression A physical act or threat of action by an individual which causes pain or injury or reduces freedom in another individual.

Aversive Causing avoidance or withdrawal.

Causal factors The inputs to the decision-making system, each of which is an interpretation of an external change or an internal state of the body. The internal state includes that of systems in the brain as well as of other body systems.

Competition (1) Among individuals, the striving of an animal to obtain for itself a resource which is in limited supply. Success might result from such abilities as speed of action, strength in fighting or ingenuity in searching. (2) Among genotypes, the ability to carry out any life function in a way which is better than that used by other genotypes so that the fitness (reproductive success) of the genotype is increased.

Cope Have control of mental and bodily stability. Lack of control may be short-lived or prolonged. Prolonged failure to be in control of mental and bodily stability leads to reduced fitness (see **Stress**).

Crowding The situation in which the movements or other activities of individuals in a group are restricted by the physical presence of others.

Displacement activity An activity which is performed in a situation which appears not to be the context in which it would normally occur. Being so dependent for recognition on observer ability to determine relevance to context, the term is of limited use.

Dominance An individual animal is said to be dominant over another when it has priority of access to a resource such as food or to a mate. A dominant individual is usually superior in fighting ability to a subordinate, but this may not have been tested.

Drive A collection of causal factors which promote related behaviours. The term often implies potential progression towards a goal. Although a definition is included here because the term 'drive' is in widespread use, it is easier to understand motivation through reference to causal factors than to drives.

Ethogram A detailed description in space and time of each behaviour shown by members of a particular species.

Ethology The observation and detailed description of behaviour in order to find out how biological mechanisms function. Sometimes such studies are carried out in a natural or semi-natural setting, but the study of animals on farms or in laboratories is also ethology.

Experience A change in the brain which results from acquiring additional information. The information can originate in the external environment of the individual or within the body, and could result from, for example, sensory input, low oxygen availability, or altered hormone levels in the blood.

Exploration Any activity which has the potential to allow an individual to acquire new information about its environment or itself.

Feedback The effect of a system output, in response to a system input, which modifies that input by reducing it (negative feedback) or enhancing it (positive feedback).

Feedforward The effect of a system output which, prior to any input, modifies the state of the system in such a way that the effect of an input is partly or wholly nullified.

Fitness Success of an individual in passing on genes to future generations. Fitness expresses lifetime reproductive success, which is reduced by mortality, delay in breeding and reduced offspring production per breeding. Many indicators reveal when fitness reduction is likely.

Flight distance The radius of space around an animal within which intrusion provokes a flight reaction.

Flight reaction A characteristic escape reaction which is specific to a particular enemy and surroundings, occurring as soon as the intruder approaches within a given distance.

Functional systems The different sorts of biological activity in the living animal which together make up the life process, such as temperature regulation, feeding and predator avoidance. These functional systems have behavioural and physiological components.

Genotype The genetic constitution of an individual organism.

Grooming The cleaning of the body surface or rearrangement of pelage by licking, nibbling, picking, rubbing, scratching, etc. When action is directed towards the animal's own body, it is called self-grooming, when directed at another individual, it is referred to as allogrooming.

Habituation The waning of an individual's response, which could still be shown, to a constant or repeated stimulus. The process is distinct from fatigue.

Hierarchy The ordering, based upon some graded ability or characteristic, of individuals or groups in a social system. The term is most frequently used where the ability assessed is that of winning fights or displacing other individuals.

Homeostasis The relatively steady state of a body variable which is maintained by means of physiological or behavioural regulation.

Individual distance The maximum distance from an animal within which approach elicits attack or avoidance. The distance may differ for approaches by different animal species or individuals.

Learning A change in the brain which results in behaviour being modified as a consequence of information acquired from outside the brain. The modification must last for longer than a few seconds otherwise the effect could simply be a reflex.

Motivation The system within the brain that induces behavioural and physiological changes, and determines which changes occur and when.

Motivational state The level of motivation resulting from the combined levels of all causal factors in the brain.

Need A requirement, fundamental in the biology of the animal, to obtain a particular resource or respond to a particular environmental or bodily stimulus. To have a 'need' is to have a deficiency which can be remedied by obtaining a specific resource or stimulus.

Overcrowding Crowding such that the fitness of individuals in the group is reduced.

Pain A sensation which, without involving higher level brain processing, such as that associated with fear, is very aversive. Pain usually involves specialized nociceptive neurons, and often involves some degree of tissue injury. Pain normally elicits protective motor and autonomic reactions, causes emotional responses, results in learned avoidance behaviour, and may modify social and other behaviour. Detection and assessment of pain in animals relies upon a combination of behavioural and physiological indices.

Periodicity The occurrence of a series of events separated by equal periods of time.

Phenotype The observable properties of an organism as they have developed under the combined influences of the genetic constitution of the individual and the effects of environmental factors.

Pheromone A substance produced by one animal which conveys information to other individuals by olfactory means.

Preening Grooming activity in birds.

Proprioceptor Sensory receptors within the body which transmit information about the relative positions of different parts of the body.

Reaction time The time between the occurrence of a stimulus and the beginning of the response of the animal.

Reinforcer An environmental change which increases the likelihood that an animal will make a particular response, i.e. a reward (positive reinforcer) or a cessation of punishment (negative reinforcer).

Rhythm A series of events repeated at approximately regular intervals.

Rights Definitions of animal rights which may be useful are: (1) A legal entitlement which can be defended using the laws of the country. In most countries animals do not have rights in this sense. (2) A privilege or entitlement justified on moral, sometimes religious, grounds. Many people find it easier to understand and support the idea of human obligations towards animals, rather than of animal rights.

Sensitive period A period during social development during which an individual's behaviour, at that time or later, is especially likely to be affected by certain types of experience.

Sensitization The increase in response to continuing or repeated stimulation.

Social facilitation Behaviour that is initiated or increased in rate or frequency by the presence of another animal carrying out the same behaviour.

Stereotypy A repeated, relatively invariate sequence of movements which has no obvious function.

Stimulation Factors affecting an individual animal or part of it.

Stimulus An environmental change which excites one or more receptors or other parts of the nervous system of an animal.

Strain The short-term consequences of stress.

Stress An environmental effect on an individual which overtaxes its control systems and reduces its fitness or appears likely to do so. Fitness reduction involves increased mortality and failure to grow or reproduce.

Territory An area which an animal defends by fighting or by demarcation. Other individuals detect the mark or other signal, which is a deterrent to entry.

Welfare The state of an individual as regards its attempts to cope with its environment.

References

Adams, C. and Rinnie, R.W. (1982) Stress protein formation: gene expression and environmental interaction with evolutionary significance. *Int. Rev. Cytol.*, **79**, 305–15.

Adams, M.R., Kaplan, J.R., Manuck, S.B. *et al.* (1988) Persistent sympathetic nervous system arousal associated with tethering in Cynomolgus macaques. *Lab. Animal Science*, **38**, 279–81.

Agger, J.F. (1983) Production disease and mortality in dairy cows; analysis of records from disposed plants from 1969–1982, in *Proc. 5th Int. Conf. Prod. Dis. Farm. Anim.*, Uppsala, pp. 308–11.

Akil, H., Madden, J., Patrick, R.L. and Barchas, J.D. (1976) Stress-induced increase in endogenous opiate peptides: concurrent analgesia and its partial reversal by naloxone, in *Opiate and Endogenous Opiate Peptides*, (ed H.W. Kosterlitz), North Holland, Amsterdam.

Akil, H., Shiomi, H. and Matthews, J. (1985) Induction of the intermediate pituitary by stress: synthesis and release of a nonopioid form of β-endorphin. *Science*, NY, **227**, 424–26.

Akil, H., Watson, S.J., Young, E. (1984) Endogenous opioids: biology and function. *Ann. Rev. Neurosci.*, **7**, 223–55.

Al-Gahtani, S.J. and Rodway, R.G. (1991) Plasma β-endorphin and cortisol in sheep during isolation stress. *Anim. Prod.*, **52**, 580.

Amit, Z. and Gallina, H. (1986) Stress-induced analgesia: adaptive pain suppression. *Physiol. Rev.*, **66**, 1091–121.

Anand, K.J.S. (1986) The stress response to surgical trauma: from physiological basis to therapeutic implications. *Prog. Food Nutr. Sci.*, **10**, 67–132.

Anand, K.J.S. and Aynsley-Green, A. (1988) Measuring the severity of surgical stress in new born infants. *J. Pediat. Surg.*, **23**, 297–305.

Anand, K.J.S., Sippell, W.G. and Aynsley-Green, A. (1987) Randomised trial of Fentanyl anaesthesia in preterm babies undergoing surgery: effects on the stress response. *Lancet*, 1243–48.

Anand, K.J.S., Sippell, W.G., Schofield, N.M. and Aynsley-Green, A. (1988) Does halothane anaesthesia decrease the metabolic and endocrine stress responses of newborn infants undergoing operation? *Br. Med. J.*, **296**, 668–77.

Anderson, I.D., Forsling, M.L., Little, R.A. and Pyman, J.A. (1989) Acute injury is a potent stimulus for vasopressin release in man. *J. Physiol.*, **416**, 28P.

Andreae, U. (1979) Zur Aktivitatsfrequenz von Mastbullen bei Spaltenbodenhaltung. *Landbauforschung Volkenrode*, **48**, 89–94.

Andreae, U. and Smidt, D. (1982) Behavioural alterations in young cattle on slatted floors, in *Disturbed Behaviour in Farm Animals*, (ed W. Bessei), Eugen Ulmer, Stuttgart.

Anil, M.H., Fordham, D. and Rodway, R. (1990) Plasma beta-endorphin increase in sheep after electrical stunning. *Br. Vet. J.*, **146**, 476–77.

Anisman, H., Pizzino, A. and Sklar, L.S. (1980) Coping with stress, norepinephrine depletion and escape performance. *Brain Res.*, **191**, 583–88.

Appleby, M.C. and Lawrence, A. B. (1987) Food restriction as a cause of stereotypic behaviour in tethered gilts. *Anim. Prod.* **45**, 103–10.

Archer, J. (1979) Behavioural aspects of fear in animals and man, in *Fear in Animals and Man*, (ed W. Sluckin), Van Nostrand Rheinhold, Princeton, New Jersey.

Arey, D.S. (1992) Straw and food as reinforcers for prepartal sows. *Appl. Anim. Behav. Sci.*, **33**, 217–26.

Armario, A., Castellanos, J.M. and Balasch, J. (1984a) Adaptation of anterior pituitary hormones to chronic noise stress in male rats. *Behav. Neur. Biol.*, **41**, 71–76.

Armario, A., Castellanos, J.M. and Balasch, J. (1984b) Dissociation between corticosterone and growth hormone adaptation to chronic stress in the rat. *Horm. Metab. Res.*, **16**, 142–45.

Armario, A. and Jolin, T. (1989) Incidence of intensity and duration of exposure to various stressors on serum TSH and GH levels in adult male rats. *Life Sci.*, **44**, 215–21.

Armario, A., Lopez-Calderon, A., Jolin, T. and Balasch, J. (1986) Response of anterior pituitary hormones to chronic stress. The specificity of adaptation. *Neurosci. Biobehav. Rev.* **10**, 245–50.

Arnold, G.W. (1985) Parturient behaviour, in *Ethology of Farm Animals*, World Animal Science A5, (ed A.F. Fraser), Elsevier, Amsterdam, pp. 335–47.

Arnold, G.W. and Estep, D.Q. (1990) Effects of housing on social preference and behaviour in male golden hamsters. *Appl. Anim. Behav. Sci.*, **27**, 253–61.

Augustini, C., Fischer, K. and Schön, L. (1977) Auswirkungen unterschiedlicher Transport belastungen auf intra vitam und post mortem enfarbare parameter beim Schwein. *Die Fleischwirtschaft*, **57**, 2037–43.

Axelrod, J. (1984) The relationship between the stress hormones, catecholamines ACTH and glucocorticoids, in *Stress: the Role of Catecholamines and other Neurotransmitters*, vol. 1, (eds E. Usdin, R. Kvetnansky and J. Axelrod), Gordon and Breach, New York, pp. 3–13.

Bacou, F. and Bressot, C. (1976) Increased plasma creatine kinase activity in rabbits: effects of systematically repeated blood sampling. *Experientia*, **32**, 487–89.

Bailey, K.J., Stephens, D.B. and Delaney, C.C. (1986) Observations on the effects of vibration and noise on plasma ACTH and zinc levels, pregnancy and respiration rate in the guinea pig. *Lab. Anim.*, **20**, 101–108.

Baker, G.H.B., Irani, M.S., Byrom, N.A. *et al.* (1985) Stress, cortisol concentrations and lymphocyte subpopulations. *Br. Med. J.*, **290**, 1393.

Baldock, N.M. and Sibly, R.M. (1990) Effects of handling and transportation on heart rate and behaviour in sheep. *Appl. Anim. Behav. Sci.*, **28**, 15–39.

Baldock, N. M., Sibly, R.M. and Penning, P.D. (1988) Behaviour and seasonal variation in heart rate in domestic sheep (*Ovi saries*). *Anim. Behav.*, **36**, 35–43.

Baldwin, B.A. (1972) Operant conditioning techniques for the study of thermoregulatory behaviour in sheep. *J. Physiol.*, **226**, 41–42P.

Baldwin, B.A. (1979) Operant studies on the behaviour of pigs and sheep in relation to the physical environment. *J. Anim. Sci.*, **49**, 1125–34.

Baldwin, B.A. and Start, I.B. (1985) Illumination preferences of pigs. *Appl. Anim. Behav. Sci.*, **14**, 233–43.

Banks, E.M. (1982) Behavioural research to answer questions about animal welfare. *J. Anim. Sci.*, **54**, 434–46.

Barclay, R.J., Herbert, W.J. and Poole, T.B. (1988) *Disturbance index method for assessing severity of procedures on rodents*, Universities Federation for Animal Welfare, Potters Bar.

Barnett, J.L., Hemsworth, P.H., Winfield, C.G. and Fahy, V.A. (1987) The effects of pregnancy and parity number on behavioural and physiological responses related to the welfare status of individual and group-housed pigs. *Appl. Anim. Behav. Sci.*, **17**, 229–43.

Barta, O. (1983) Serum's lymphocyte immunoregulatory factors (SLIF). *Vet. Immunol. Immunopathol.*, **4**, 279–306.

Bartrop, R.W., Luckhurst, E., Lazarus, L. Kiloh, L.G. and Penny, R. (1977) Depressed lymphocyte function after bereavement. *Lancet*, **1**, 834–36.

Bateson, P. (1986) Functional approaches to behaviour and development, in *Primate Ontogeny, Cognition and Social Behaviour,* (eds J.G. Else and P.C. Lee), Cambridge University Press, Cambridge, pp. 183–92.

Bateson, P. (1991) Assessment of pain in animals. *Anim. Behav.*, **42**, 827–39.

Baum, A., Grunberg, N.E. and Singer, J.E. (1982) The use of psychological and neuroendocrinological measurements in the study of stress. *Health Psychol.*, **1**, 1217–36.

Baxter, M.R. (1988) Needs – behavioural or psychological? *Appl. Anim. Behav. Sci.*, **19**, 345–48.

Becker, B.A. (1987) The phenomenon of stress: Concepts and mechanisms associated with stress-induced responses of the neuroendocrine system. *Vet. Res. Comm.*, **11**, 443–56.

Beilharz, R. G. (1985) Special phenomena, in *Ethology of Farm Animals*, World Animal Science A5, (ed A.F. Fraser), Elsevier, Amsterdam.

Benus, I. (1988) Aggression and coping. Differences in behavioural strategies between aggressive and non-aggressive male mice. Ph.D. thesis, University of Groningen.

Blackshaw, J. K. and McVeigh, J. F. (1984) The behaviour of sows and gilts, housed in stalls, tethers and groups. *Proc. Aust. Soc. Anim. Prod.*, **15**, 85–88.

Blaese, R.M., Weiden, P., Oppenheim, J.J. and Waldmann, T.A. (1973) Phytohaemagglutinin as a skin test for the evaluation of cellular immune competence in man. *J. Lab. Clin. Med.*, **81**, 538–48.

Blanchard, S.C. and Chang, K.J. (1988) Regulation of opioid receptors, in *The Opiate Receptors* (ed G.W. Pasternak), The Humana Press, Clifton, New Jersey, pp. 430–39.

Blecha, F., Boyles, S.L. and Riley, J.G. (1984) Shipping suppresses lymphocyte blastogenic response in Angus and Brahman × Angus feeder calves. *J. Anim. Sci*, **59**, 576–83.

Bligh, J. (1973) *Temperature Regulation in Mammals and other Vertebrates*, North-Holland, Amsterdam, p. 167.

Block, W. (1985) Survival on land. *Biologist*, **32,** 132–38.

Bodnar, R.J. (1984) Types of stress which induce analgesia, in *Stress-Induced Analgesia*, (eds M.D. Tricklebank and G. Curzon), Wiley, Chichester, pp. 19–32.

Boer, S.F. de, Koopmans, S.J., Slangen, J.L. and Van der Gugten, J. (1989a) Effects of fasting on plasma catecholamine, corticosterone and glucose concentrations under basal and stress conditions in individual rats. *Physiol. Behav.*, **45**, 989–94.

Boer, S.F. de, van der Gugten, J. and Slangen, J.L. (1989b) Plasma catecholamine and corticosterone responses to predictable and unpredictable noise stress in rats. *Physiol. Behav.*, **45**, 789–95.

Bohus, B. (1974) Telemetered heart rate responses of the rat during free and learned behavior. *Biotelemetry*, **1**, 193–201.

Bohus, B., Menus, R.F. and Fokkema, D.S. et al. (1987) Neuroendocrine states and behavioral and physiological stress responses. *Prog. Brain. Res.*, **72,** 57–70.

Bold, A.J. de (1985) Atrial natriuretic factor: a hormone produced by the heart. *Science*, NY, **230,** 767–70.

Bonnet, K.A., Hiller, J.M. and Simon, E.J. (1976) The effects of chronic opiate treatment and social isolation on opiate receptors in rodent brain, in *Opiates and Endogenous Opioid Peptides*, (ed H.W. Kosterlitz), Elsevier, Amsterdam, pp. 335–43.

Brambell, F.W.R. (1965) *Report on the Technical Committee to enquire into the welfare of livestock kept under intensive husbandry conditions*, HMSO, London.

Breazile, J.E. (1988) The physiology of stress and its relationship to mechanisms of disease and therapeutics. *Vet. Clinics N. America: Food Animal Practice*, **4.3**, 441–80.

Brenner, B.M., Ballermann, B.J., Gunning, M.E. and Zeidel, M.L. (1990) Diverse biological actions of atrial natriuretic peptide. *Physiol. Rev.*, **70**, 665–99.

Brion, A. (1964) Les tics chez les animaux, in *Psychiatrie Animale*, (eds A. Brion and H. Ey), Desclée de Brouwer, Paris, pp. 299–306.

Brooks, J.E., Herbert, M., Walder, C.P. *et al.* (1986) Prolactin and stress: some endocrine correlates of pre-operative anxiety. *Clin. Endocrinol.*, **24**, 653-56.

Broom, D.M. (1968) Specific habituation by chicks. *Nature, Lond.*, **217**, 880–81.

Broom, D.M. (1969a) Effects of visual complexity during rearing on chicks' reactions to environmental change. *Anim. Behav.*, **17**, 773–80.

Broom, D.M. (1969b) Reactions of chicks to visual changes during the first ten days after hatching. *Anim. Behav.*, **17**, 307–15.

Broom, D.M. (1981a) Behavioural plasticity in developing animals, in *Development in the Nervous System*, (ed D.R. Garrod and J.D. Feldman), Cambridge University Press, Cambridge, pp. 361–78.

Broom, D.M. (1981b) *Biology of Behaviour*, Cambridge University Press, Cambridge.

Broom, D.M. (1982) Husbandry methods leading to inadequate social and maternal behaviour in cattle, in *Disturbed Behaviour in Farm Animals*, (ed W. Bessei), Eugen Ulmer, Stuttgart.

Broom, D.M. (1983a) Stereotypies as animal welfare indicators, in *Indicators Relevant to Farm Animal Welfare*, (ed D. Smidt), *Curr. Top. Vet. Med. Anim. Sci.*, Martinus Nijhoff, The Hague, pp. 81–87.

Broom, D.M. (1983b) The stress concept and ways of assessing the effects of stress in farm animals. *Appl. Anim. Ethol.*, **1**, 79.

Broom, D.M. (1985) Stress, welfare and the state of equilibrium, in *Proc. 2nd Eur. Symp. Poult. Welfare*, (ed R.M. Wegner), World Poultry Science Association, Celle, pp. 72–81.

Broom, D.M. (ed) (1986a) *Farmed Animals*, Torstar Books, New York.

Broom, D.M. (1986b) Indicators of poor welfare. *Br. Vet. J.*, **142**, 524–26.

Broom, D.M. (1987) Applications of neurobiological studies to farm animal welfare, in *Biology of Stress in Farm Animals: an integrated approach*, (eds

P.R. Wiepkema and P.W.M. van Adrichem), *Curr. Top. Vet. Med. Anim. Sci.*, Martinus Nijhoff, Dordrecht, pp. 101–10.

Broom, D.M. (1988a) Needs, freedoms and the assessment of welfare. *Appl. Anim. Behav. Sci.*, **19**, 384–86.

Broom, D.M. (1988b) The scientific assessment of animal welfare. *Appl. Anim. Behav. Sci.*, **20**, 5–19.

Broom, D.M. (1988c) Les concepts de stress et de bien-être. *Rec. Méd. vét.*, **164**, 715–22.

Broom, D.M. (1988d) The relationship between welfare and disease susceptibility in farm animals, in *Animal Disease – a Welfare Problem*, (ed T.E. Gibson), BVA Animal Welfare Foundation, London, pp. 22–29.

Broom, D.M. (1990) The importance of measures of poor welfare. *Behav. Brain Sci.*, **13**, 14.

Broom, D.M. (1991a) Animal welfare: concepts and measurement. *J. Anim. Sci.*, **69**, 4167–75.

Broom, D.M. (1991b) Assessing welfare and suffering. *Behav. Process.*, **25**, 117–23.

Broom, D.M. (1991c) Needs and welfare of housed calves, in *New Trends in Veal Calf Production*, (eds J.H.M. Metz and C.M. Groenestein), Pudoc, Wageningen, pp. 23–31.

Broom, D.M. (1992) The needs of laying hens and some indicators of poor welfare, in *The Laying Hen*, (eds V. and H. Carter), European Conference Group on the Protection of Farm Animals, Horsham, pp. 4–19.

Broom, D.M. (in press) A usable definition of animal welfare. *J. Agric Ethics.*

Broom, D.M., Knight, P.G. and Stansfield, S.C. (1986) Hen behaviour and hypothalamic pituitary-adrenal responses to handling and transport. *Appl. Anim. Behav. Sci.*, **16**, 98.

Broom, D.M. and Leaver, J.D. (1977) Mother–young interactions in dairy cattle. *Br. Vet. J.*, **133**, 192.

Broom, D.M. and Leaver, J.D. (1978) The effects of group-housing or partial isolation on later social behaviour of calves. *Anim. Behav.*, **26**, 1255–63.

Broom, D.M. and Potter, M.J. (1984) Factors affecting the occurrence of stereotypies in stallhoused dry sows, in *Proc. Int. Cong. Appl. Ethol. Farm Anim.*, (eds J. Unshelm, G. van Putten and K. Zeeb), KTBL, Darmstadt, pp. 229–31.

Brown, A.G. (1991) *Nerve Cells and Nervous Systems: an introduction to neuroscience*, Springer-Verlag, London.

Budiansky, S. (1992) *The Covenant of the Wild: why animals chose domestication*, William Morrow Inc., New York.

Cabanac, M., and Johnson, K.G. (1983) Analysis of a conflict between palatability and cold exposure in rats. *Physiol. Behav.*, **31**, 249–53.

Cairns, R.B., Hood, K.E. and Midlam, J. (1985) On fighting in mice: is there a sensitive period for isolation effects? *Anim. Behav.*, **33**, 166–80.

Calabrese, J.R., Kling, M.A. and Gold, P.W. (1987) Alterations in immunocompetence during stress, bereavement and depression: focus on neuroendocrine regulation. *Am. J. Psychiat.*, **144**, 1123–34.

Calabrese, J.R., Skwerer, R.G., Barna, B., Gulledge, A.D., Valenzuela, R., Buktus, A., Subichin, S. and Krupp, N.E. (1986) Depression, immunocompetence and prostaglandins of the E series. *Psychiat. Res.*, **17**, 41–47.

Calhoun, J.B. (1962) *The Ecology and Sociology of the Norway Rat*, US Dept. Health, Educ. Welf. PHS Doc. 1008, US Govt. Printing Office, Washington, D.C.

Cancela, L.M., Artinián, J., and Fulginiti, S. (1988) Opioid influence on some aspects of stereotyped behaviour induced by repeated amphetamine treatment. *Pharmacol. Biochem. Behav.*, **30**, 899–904.

Cannon, W.B. (1935) Stresses and strains of homeostasis. *American Journal of the Medical Sciences*, **189**, 1–14.

Carlstead, K. (1986) Predictability of feeding: its effect on agonistic behaviour and growth in grower pigs. *Appl. Anim. Behav. Sci.*, **16**, 25.

Castagné, V., Corder, R., Gaillard, R. and Mormède, P. (1987) Stress-induced changes of circulating neuropeptide Y in the rat: comparison with catecholamines. *Regul. Pept.*, **19**, 55–63.

Chamove, A.S. (1989a) Cage design reduces emotionality in mice. *Lab. Anim.*, **23**, 215–19.

Chamove, A.S. (1989b) Environmental enrichment: a review. *Anim. Technol.*, **40**, 155–77.

Chamove, A.S., Rosenblum, L.A., and Harlow, H.F. (1973) Monkeys (*Macaca mulatta*) raised only with peers. A pilot study. *Anim. Behav.*, **21**, 316–25.

Charlesworth, B. (1980) *Evolution in Age-Structured Populations*, Cambridge University Press, Cambridge.

Charlton, B. (1991) Stress? Who needs it? *New Scientist*, June, p. 55.

Charrière, H. (1969) *Papillon*, Robert Laffort, Paris.

Christian, J.J. (1961) Phenomena associated with population density. *Proc. Natn. Acad. Sci., USA*, **47**, 428–91.

Clark, R.B. (1960) Habituation of the polychaete *Nereis* to sudden stimuli, 1. General properties of the habituation process. *Anim. Behav.*, **8**, 82–91.

Coe, C.L., Lubach, G.R., Ershler, W.B. and Klopp, R.G. (1989) Influence of early rearing on lymphocyte proliferation responses in juvenile Rhesus monkeys. *Brain. Behav. Immunol.*, **3**, 47–60.

Coe, C.L., Rosenberg, L.T. and Levine, S. (1988) Effect of maternal separation on the complement system and antibody responses in infant primates. *Int. J. Neurosci.*, **40**, 289–302.

Cohen, M.R., Cohen, R.M., Dickor, D. *et al.* (1983) High dose naloxone infusions in normals. *Arch. Gen. Psych.*, **40**, 613–19.

Cohen, M., Pickar, D., Dubois, M. *et al.* (1981) Surgical stress and endorphins. *Lancet*, **1**, 213–14.

Colgan, P. (1989) *Animal Motivation*, Chapman & Hall, London.

Cooper, C.L. and Payne, R. (eds) (1988) *Causes, Coping and Consequences of Stress at Work*, Wiley, Chichester.

Cooper, B.Y. and Vierck, C.J. (1986) Measurement of pain and morphine hypoalgesia in monkeys. *Pain*, **26**, 361–92.

Cooper, T.R., Trunkfield, H.R., Zanella, A.J. and Booth, W.D. (1989) An enzyme-linked immunosorbent assay for cortisol in the saliva of man and domestic farm animals. *J. Endocrinol.*, **123**, R13–R16.

Corley, K.C., Shiel, F. O'M., Mauck, H.P. and Greenhoot, J. (1973) Electrocardiographic and cardiac morphological changes associated with environmental stress in squirrel monkeys. *Psychosom. Med.*, **35**, 361–64.

Craig, J.V., Craig, J.A. and Vargas Vargas, J. (1986) Corticosteroids and other indicators of hens' well-being in four laying-house environments. *Poult. Sci.*, **65**, 856–63.

Cronin, G.M., Wiepkema, P.R and van Ree, J.M. (1985) Endogenous opioids are involved in abnormal stereotyped behaviours of tethered sows. *Neuropeptides*, **6**, 527–30.

Cronin, G.M. *et al.* (1986) Endorphins implicated in stereotypies of tethered sows. *Experientia*, **42**, 198–99.

Curtis, S.E. (1983) Perception of thermal comfort by farm animals, in *Farm Animal Housing and Welfare*, (eds S.H. Baxter, M.R. Baxter and J.A.C. MacCormack), *Curr. Top. Vet. Med. Anim. Sci.*, Martinus Nijhoff, The Hague, pp. 59–66.

Dantzer, R. (1986) Behavioral, physiological and functional aspects of stereotyped behavior: a review and a reinterpretation. *J. Anim. Sci.*, **62**, 1776–86.

Dantzer, R. and Mormède, P. (1981) Pituitary adrenal consequences of adjunctive behaviours in pigs. *Horm. Behav.*, **15**, 386–95.

Dantzer, R. and Mormède, P. (1983) Stress in farm animals: a need for re-evaluation. *J. Anim. Sci.*, **57**, 6–18.

Dantzer, R. and Mormède, P. (1985) Stress in Domestic Animals: A psychoneuroendocrine approach, in *Animal Stress*, (ed G.P. Moberg), Am. Physiol. Soc., Bethesda, Maryland, pp. 81–95.

Dantzer, R., Mormède, P., Bluthe, R.-M. and Soissons, J. (1983) The effect of different housing conditions on behavioural and adrenocortical reactions in veal calves. *Reprod. Nutr. Dévelop.*, **23**, 67–74.

Dawkins, M. (1976) Towards an objective method of assessing welfare in domestic fowl. *Appl. Anim. Ethol.*, **2**, 245–54.

Dawkins, M. (1977) Do hens suffer in battery cages? Environmental preferences and welfare. *Anim. Behav.*, **25**, 1034–46.

Dawkins, M. (1981) Priorities in the cage size and flooring preferences of domestic hens. *Br. Poult. Sci.*, **22**, 255–63.

Dawkins, M. (1983) Battery hens name their price: consumer demand theory and the measurement of animal needs. *Anim. Behav.*, **31**, 1195–1205.

Dawkins, M.S. (1980) *Animal Suffering: the science of animal welfare*, Chapman & Hall, London.

Dawkins, M.S. (1988) Behavioural deprivation: a central problem in animal welfare. *Appl. Anim. Behav. Sci.*, **20**, 209–25.

Dawkins, M.S. (1990) From an animals point of view: motivation, fitness, and animal welfare. *Behav. Brain Sci.*, **13**, 1–61.

Devor, M. (1984) Pain and 'state' induced analgesia: an introduction, in *Stress-Induced Analgesia*, (eds M.D. Tricklebank and G. Curzon), Wiley, Chichester, pp. 1–18.

Dimond, S.J. and Adam, J.H. (1972) Approach behaviour and embryonic visual experience in chicks: studies on the effect of rate of visual flicker. *Anim. Behav.*, **20**, 413–20.

Dubois, M., Pickar, D., Cohen, M.R., Roth, Y.F., MacNamara, T. and Bunnay, W.E. (1981) Surgical stress in humans is accompanied by an increase in plasma Beta-endorphin immunoreactivity. *Life Sci.*, **29**, 1249–54.

Duncan, I.J.H. (1978) The interpretation of preference tests in animal behaviour. *Appl. Anim. Ethol.*, **4**, 197–200.

Duncan, I.J.H. (1986) Some thoughts on the stressfulness of harvesting broilers. *Appl. Anim. Behav. Sci.*, **16**, 97.

Duncan, I.J.H. and Filshie, J.H. (1979) The use of radiotelemetry devices to measure temperature and heart rate in domestic fowl, in *A Handbook on Biotelemetry and Radio Tracking*, (eds C.J. Amlaner and D. McDonald), Pergamon, Oxford, pp. 579–88.

Duncan, I.J.H. and Kite, V.G. (1987) Some investigations into motivation in the domestic fowl. *Appl. Anim. Behav. Sci.*, **18**, 387–88.

Duncan, I.J.H. and Molony, V. (1986) Assessment of pain using conditioning procedures, in *Assessment of Pain in Farm Animals*, (eds I.J.H. Duncan and V. Molony), Office for Official Publications of the European Communities, Luxembourg, pp. 71–75.

Duncan, I.J.H. and Petherick, J.C. (1991) The implications of cognitive process for animal welfare. *J. Anim. Sci.*, **69**, 5017–22.

Duncan, I.J.H. and Wood-Gush, D.G.M. (1971) Frustration and aggression in the domestic fowl. *Anim. Behav.*, **19**, 500–504.

Duncan, I.J.H. and Wood-Gush, D.G.M. (1972) Thwarting of feeding behaviour in the domestic fowl. *Anim. Behav.*, **20**, 444–51.

Eberhart, J.A., Keverne, E.B. and Meller, R.E. (1983) Social influences on circulating levels of cortisol and prolactin in male talapoin monkeys. *Physiol. Behav.*, **30**, 361–69.

Edens, F.W. 1987 Manifestations of social stress in grouped Japanese quail. *Comp. Biochem. Physiol.*, **86A**, 469–72.

Edmunds, M. (1974) *Defence in Animals*, Longman, Harlow.

Ekesbo, I. (1981) Some aspects of sow health and housing, in *Welfare of Pigs*, (ed W. Sybesma), *Curr. Top. Vet. Med. Anim. Sci.*, **11**, 250–266. Martinus Nijhoff, The Hague.

El-Halawani, M.E., Waibel, P.E., Appel, J.R. and Good, A.L. (1973) Effects of temperature stress on catecholamines and corticosterone of male turkeys. *Am. J. Physiol.*, **224**, 384–88.

Ely, D.L. and Henry, J.P. (1971) Effects of social role upon the blood pressure of individual male mice. *Fed. Proc.*, **30**, 265.

Engel, G.L. (1967) The psychological setting of somatic disease: the 'giving up-given up' complex. *Proc. Roy. Soc. Med.*, **60**, 553–55.

Esterling, B. and Rabin, B.S. (1987) Stress-induced alteration of T-lymphocyte subsets and humoral immunity in mice. *Behav. Neurosci.*, **101**, 115–19.

Euker, J.S., Meites, J. and Riegle, G.D. (1975) Effects of acute stress on serum LH and prolactin in intact, castrate and dexamethasone-treated male rats. *Endocrinology*, **96**, 85–92.

Evans, S.M. (1965) Learning in the polychaete *Nereis*. *Nature, Lond.*, **207**, 1420.

Ewbank, R. (1973) Use and abuse of the term 'stress' in husbandry and welfare. *Vet. Rec.*, **92**, 709–10.

Fazlul Haque, A.K.M. and Broom, D.M. (1985) Experiments comparing the use of kites and gas bangers to protect crops from woodpigeon damage. *Agric. Ecosystems Environ.*, **2**, 219–28.

Fell, L.R. and Shutt, D.A. (1986) Use of salivary cortisol as an indicator of stress due to management practices in sheep and calves. *Proc. Aust. Anim. Prod.*, **16**, 203–206.

Fell, L.R. and Shutt, D.A. (1989) Behavioural and hormonal responses to acute surgical stress in sheep. *Appl. Anim. Behav. Sci.*, **22**, 283–94.

Fell, L.R., Shutt, D.A. and Bentley, C.J. (1985) Development of a salivary cortisol method for detecting changes in plasma 'free' cortisol arising from acute stress in sheep. *Aust. Vet. J.*, **62**, 403–406.

Flecknell, P.A. (1985a) The management of post-operative pain and distress in experimental animals. *Anim. Technol.*, **36**, 97–103.

Flecknell, P.A. (1985b) Recognition and alleviation of pain in animals, in *Advances in Animal Welfare Science*, (eds M.W. Fox and L.D. Mickley), Martinus Nijhoff, Dordrecht, pp. 61–77.

Flecknell, P.A., Kirk, A.J.B., Liles, J.H., Hayes, P.H. and Darke, J.H. (1991) Post-operative analgesia following thoracotomy in the dog: an evaluation of the effects of bupivacaine intercostal nerve block and nalbuphine on respiratory function. *Lab. Anim.*, **25**, 319–24.

Flecknell, P.A. and Liles, J.H. (1991) The effects of surgical procedures, halothane anaesthesia and nalbuphine on locomotor activity and food and water consumption in rats. *Lab. Anim.*, **25**, 50–60.

Fletcher, B.C. (1991) *Work, Stress, Disease and Life Expectancy*, Wiley, Chichester.

Fordham, D.P., Lincoln, G.A., Ssewannyana, E. and Rodary, R.G. (1989) Plasma β-endorphin and cortisol concentrations in lambs after handling, transport and slaughter. *Anim. Prod.*, **49**, 103–108.

Forrester, R.C. (1979) Behavioural State and Responsiveness in Domestic Chicks. Ph.D. thesis, University of Reading.

Frädrich, H. (1967) *Handbuh der Zoologie*, **8**, 1–44.

Frankenhaeuser, M. (1975) Experimental approaches to the study of catecholamines and emotion, in *Emotions – their parameters and measurement*, (ed L. Levi), Raven, New York, pp. 209–34.

Fraser, A.F. (1960) The influence of psychological and other factors on reaction time in bulls. *Cornell Vet.*, **50**, 126–32.

Fraser, A.F. and Broom, D.M. (1990) *Farm Animal Behaviour and Welfare*, Baillière Tindall, London.

Fraser, D. (1975) The effect of straw on the behaviour of sows in tether stalls. *Anim. Prod.*, **21**, 59–68.

Fraser, D., Ritchie, J.S.D. and Fraser, A.F. (1975) The term 'stress' in a veterinary context. *Br. Vet. J.*, **131**, 653–62.

Freeman, B.M. (1971) Stress and the domestic fowl: a physiological appraisal. *Wld. Poult. Sci. J.*, **27**, 263–75.

Freeman, B.M. (1987) The stress syndrome. *Wld. Poult. Sci. J.*, **43**, 15–19.

Freeman, B.H. and Flack, I.M. (1980) Effects of handling on plasma corticosterone concentrations in the immature domestic fowl. *Comp. Biochem. Physiol.*, **66A**, 77–81.

Freeman, B.M., Kettlewell, P.J., Manning, A.G.C. and Berry, P.S. (1984) Stress of transportation for Broilers. *Vet. Rec.*, **114**, 286–87.

Freeman, B.M. and Manning, A.C.C. (1976) Mediation of glucagon in the response of the domestic fowl to stress. *Comp. Biochem. Physiol.*, **53A**, 169–71.

Frese, M. and Zapf, D. (1988) Methodological issues in the study of work stress: objective vs subjective measurement of work stress and the question of

longitudinal studies, in *Causes, Coping and Consequences of Stress at Work*, (eds C.L. Cooper and R. Payne), Wiley, Chichester, p. 376.
Friederich, M.W., Friederich, D.P. and Walker, J.M. (1987) Effects of Dynorphin (1–8) on movement; non-opiate effects and structure activity relationship. *Peptides*, **8**, 837–40.
Friend, T.H., Dellmeier, G.R. and Gbur, E.E. (1985) Comparison of four methods of calf confinement, 1. Physiology. *J. Anim. Sci.*, **60**, 1095–101.
Friend, T.H., Polan, C.E., Gwazdauskas, F.C. and Heald, C.W. (1977) Adrenal glucocorticoid response to exogenous adrenocorticotropin mediated by density and social disruption in lactating cows. *J. Dairy Sci.*, **60**, 1958–63.
Frisch, J.E. (1981) Changes occurring in cattle as a consequence of selection for growth rate in a stressful environment. *J. Agric. Sci., Camb.*, **96**, 23–38.
Frölich, M., Walma, S.T. and Souverijn, J.H.M. (1981) Probable influence of cage design on muscle metabolism of rats. *Lab. Animal Science*, **31**, 510–12.
Gabrielsen, G.W., Kanwisher, J.W. and Steen, J.B. (1977) Emotional bradycardia: a telemetry study on incubating willow grouse *Lagopus lagopus*. *Acta Physiol. Scand.*, **100**, 255–57.
Gaillard, R.-C. and Al-Damluji, S. (1987) Stress and the pituitary-adrenal axis. *Baillière's Clin. Endocr. Metab.*, **1.2**, 319–54.
Gamallo, A., Villanua, A., Trancho, G. and Fraile, A. (1986) Stress adaptation and adrenal activity in isolated and crowded rats. *Physiol. Behav.*, **36**, 217–21.
Ganong, W.F. (1987) *Review of Medical Physiology*, 9th edn, Lange Medical Publ.
Gärtner, K., Büttner, D., Döhler, K., Friedel, R., Lindena, J. and Trautschold, l. (1980) Stress response of rats to handling and experimental procedures. *Lab. Anim.*, **14**, 267–74.
Georgiev, J. (1978) Influence of environmental conditions and handling on the temperature rhythm of the rat. *Biotelemetry Patient Monit.*, **5**, 229–34.
Gibbs, D.M. (1984) Dissociation of oxytocin,vasopressin and corticotropin secretion during different types of stress. *Life Sci.*, **35**, 487–91.
Gibbs, D.M. (1986a) Stress-specific modulation of ACTH secretion by oxytocin. *Neuroendocrinol.*, **42**, 456–58.
Gibbs, D.M. (1986b) Vasopressin and oxytocin: hypothalamic modulators of the stress response: a review. *Psychoneuroendocrinol.*, **11**, 131–40.
Giraldi, T., Perissin, L., Zorzet, S., Piccini, P. and Rapozzi, V. (1989) Effects of stress on tumour growth and metastasis in mice bearing Lewis lung carcinoma. *Eur. J. Cancer Clin. Oncol.*, **25**, 1583–88.
Glaser, R., Kiecolt-Glaser, J.K., Stout, J.C., Tarr, K.L., Speicher, C.E. and Holliday, J.E. (1985) Stress-related impairments in cellular immunity. *Psychiatry Res.*, **16**, 233–39.
Glaser, R., Mehl, V.S., Penn, G., Speicher, C.E. and Kiecolt-Glaser, J.K. (1986a) Stress-associated changes in plasma immunoglobulin levels. *Int. J. Psychosom.*, **33**, 41–42.
Glaser, R., Rice, J., Stout, J.L., Speicher, C.E. and Kiecolt-Glaser, K.A. (1986b) Stress depresses interferon production by leukocytes concomitant with a decrease in natural killer cell activity. *Behav. Neurosci.*, **100**, 675–78.
Goldsmith, J.F., Brain, P.F. and Benton, D. (1976) Effects of age of differential housing and duration of individual housing/grouping on intermale fighting behavior and adrenocortical activity in T.O. strain mice. *Aggressive Behav.*, **2**, 307–23.

Goldstein, D.S. (1987) Stress-induced activation of the sympathetic nervous system. *Baillière's Clin. Endocr. Metab.*, **1.2**, 253–78.
Gomez, R.E., Büttner, D. and Cannota, M.A. (1989) Open field behaviour and cardiomuscular response to stress in normal rats. *Physiol. Behav.*, **45**, 767–69.
Goncharov, N.P., Tanarov, A.G., Antonichev, A.V., Gorlushkin, V.M., Aso, T., Cekan, S.Z. and Diczfalusy, E. (1979) Effect of stress on the profile of plasma steroids in baboons (*Papio hamadyras*). *Acta Endocrinol.*, **90**, 372–84.
Goncharov, N.P., Tavadyan, D.S., Powell, J.E. and Stevens, V.C. (1984) Levels of adrenal and gonadal hormones in Rhesus monkeys during chronic hypokinesia. *Endocrinology*, **115**, 129–35.
Gregory, N.G. and Wilkins, L.J. (1989) Broken bones in chickens, I. Handling and processing damage in end of lay battery hens. *Br. Poult. Sci.*, **30**, 555–62.
Griffin, D.R. (1981) *The Question of Animal Awareness*, Rockefeller University Press, New York.
Griffin, J.F.T. (1989) Stress and immunity: a unifying concept. *Vet. Immunol. Immunopath.*, **20**, 263–312.
Gross, W.B. (1962) Blood cultures, blood counts and temperature records in an experimentally produced 'air sac disease' and uncomplicated *Escherichia coli* infection of chickens. *Poult. Sci.*, **41**, 691–700.
Gross, W.B. and Colmano, G. (1965) The effect of social isolation on resistance to some infectious diseases. *Poult. Sci.*, **48**, 515–20.
Gross, W.B. and Siegel, P.B. (1975) Immune response to *Escherichia coli*. *Am. J. Vet. Res.*, **36**, 568–71.
Gross, W.B and Siegel, P.B. (1981) Long term exposure of chickens to three levels of social stress. *Avion Dis.*, **25**, 312–25.
Grossman, A. and Rees, L.H. (1983) The neuroendocrinology of opioid peptides. *Br. Med. Bull.*, **39**, 83–88.
Groves, P.M. and Thompson, R.F. (1970) Habituation: a dual process theory. *Psychol. Rev.*, **77**, 419–50.
Guillemin, R., Vargo, T., Rossier, J., Minick, S., Ling, N., Rivier, C., Vale, W. and Bloom, F. (1977) β-endorphin and adrenocorticotropin are secreted concomitantly by the pituitary gland. *Science*, NY, **197**, 1367–69.
Guise, H.J. and Penny, R.H.C. (1989a) Factors influencing the welfare and carcass and meat quality of pigs, 1. The effects of stocking density in transport and the use of electric goads. *Anim. Prod.*, **49**, 511–15.
Guise, H.J. and Penny, R.H.C. (1989b) Factors influencing the welfare and carcass and meat quality of pigs, 2. Mixing unfamiliar pigs. *Anim. Prod.*, **49**, 517–21.
Guyton, A.C. (1991) *Textbook of Medical Physiology*, 8th edn, W.B. Saunders, Philadelphia.
Hails, M.R. (1978) Transport stress in animals: a review. *Anim. Regul. Stud.*, **1**, 289–343.
Halliday, T.R. and Sweatman, H.P.A. (1976) To breathe or not to breathe: the newt's problem. *Anim. Behav.*, **24**, 551–61.
Hanin, I., Frazer, A., Croughan, J., Davis, J.M., Katz, M.M., Koslow, S.H., Maas, J.W. and Stokes, P.E. (1985) Depression: implications of clinical studies for basic research. *Fed. Proc.*, **44**, 85–90.
Harlow, H.F. and Harlow, M.K. (1965) The affectional systems, in *Behavior of Nonhuman Primates*, vol. 2, (eds A.M. Schrier, H.F. Harlow and F. Stollnitz), Academic Press, New York.

Harrison, R. (1964) *Animal Machines*, Stuart, London.
Hayashi, K.T. and Moberg, G.P. (1987) Influence of acute stress and the adrenal axis on regulation of LH and testosterone in the male Rhesus monkey. *Am. J. Primatol.*, **12**, 263–73.
Hediger, H. (1934) Über bewegungstereotypien beim gehaltenen Tieren, *Rev. suisse Zool.*, **42**, 349–56.
Hediger, H. (1941) Biologische Gestzmässigkeiten im Verhalten von Wirbeltieren. *Mitt. Naturf. Ges. Bern.*
Heller, K.E., Houbak, B. and Jeppesen, L.L. (1988) Stress during mother–infant separation in ranch mink. *Behav. Process.*, **17**, 217–77.
Heller, K.E. and Jeppesen, L.L. (1985) Behavioural and eosinophil leukocyte responses to single and repeated immobility stress in mink. *Scientifur.*, **9**, 174–78.
Hemmer, H. (1990) *Domestication: the decline of environmental appreciation*, 2nd edn, Cambridge University Press, Cambridge.
Hemsworth, P.H. and Beilharz, R.G. (1979) The influence of restricted physical contact with pigs during rearing on the sexual behaviour of the male domestic pig. *Anim. Prod.*, **29**, 311–14.
Hemsworth, P.H., Barnett, J.L. and Hansen, C. (1986) The influence of early contact with humans and subsequent behavioural responses of pigs to humans. *Appl. Anim. Behav. Sci.*, **15**, 55–63.
Hemsworth, P.H., Barnett, J.L. and Hansen, C. (1987) The influence of inconsistent handling by humans on the behaviour, growth and corticosteroids of young pigs. *Appl. Anim. Behav. Sci.*, **17**, 245–52.
Hemsworth, P.H., Brand, A, and Willens, P.J. (1981) The behavioural response of sows to the presence of human beings and their productivity. *Livestock Prod. Sci.*, **8**, 67–74.
Hemsworth, P.H., Findlay, J.K. and Beilharz, R.G. (1978) The importance of physical contact with other pigs during rearing on the sexual behaviour of the male domestic pig. *Anim. Prod.*, **27**, 201–207.
Hennessy, M.B. (1986) Multiple, brief maternal separations in the Squirrel monkey: changes in hormonal and behavioral responsiveness. *Physiol. Behav.*, **36**, 245–50.
Hennessy, M.B., Heybach, J.P., Vernikos, J. and Levine, S. (1979) Plasma corticosterone concentrations sensitively reflect levels of stimulus intensity in the rat. *Physiol. Behav.*, **22**, 821–25.
Hennessy, M.B. and Levine, S. (1978) Sensitive pituitary-adrenal responsiveness to varying intensities of psychological stimulation. *Physiol. Behav.*, **21**, 297.
Henry, J.P. (1976) Mechanisms of psychosomatic disease in animals. *Adv. Vet. Sci. Comp. Med.*, **20**, 115–45.
Henry, J.P., Ely, D.L., Stephens, P.M., Ratcliffe, H.L. and Santisteban, G.A. (1971a) The role of psychosocial factors in the development of arteriosclerosis in CBA mice. *Atherosclerosis.*, **14**, 203–18.
Henry, J.P. and Stephens, P.M. (1977a) Endocrine characteristics of dominant and subordinate status in animals, in *Stress, Health and the Social Environment*, (eds J.P. Henry and P.M. Stephens), Springer-Verlag, New York, pp. 131–34.
Henry, J.P. and Stephens, P.M. (1977b) Monitoring behavioural disturbances in experimental social systems, in *Stress, Health and the Social Environment*, (eds J.P. Henry and P.M. Stephens), Springer-Verlag, New York, pp. 69–91.

Henry, J.P., Stephens, P.M., Axelrod, J. and Mueller, R.A. (1971b) Effect of psychosocial stimulation on the enzymes involved in the biosynthesis and metabolism of noradrenaline and adrenaline. *Psychosom. Med.*, **33**, 227–37.

Henry, J.P., Stephens, P.M. and Santisteban, G.A. (1975) A model of psychosocial hypertension showing reversibility and progression of cardiovascular complications. *Circ. Res.*, **36**, 156–64.

Herd, J.A., Morse, W.H., Kelleher, R.T. and Jones, L.G. (1969) Arterial hypertension in the squirrel monkey during behavioral experiments. *Am. J. Physiol.*, **217**, 24–29.

Hiramatsu, M., Nabeshima, T., Furukawa, H. and Kameyama, T. (1987) Different effects of ethylketocyclazocine on phencyclidine and N-allylnormetazocine-induced stereotyped behaviour in rats. *Pharmacol. Biochem. Behav.*, **28**, 489–94.

Hobfoll, S.E. (1989) Conservation of resources: a new attempt at conceptualizing stress. *Amer. Psychol.*, **44**, 513–24.

Hofer, M.A. (1970) Cardiac and respiratory function during sudden prolonged immobility in wild rodents. *Psychosom. Med.*, **32**, 633–47.

Holst, D. von (1986) Vegetative and somatic components of tree shrews' behaviour. *J. Auton. Nerv. Syst. Suppl.*, 657–70.

Horn, G. (1967) Neuronal mechanisms of habituation. *Nature, Lond.*, **215**, 707–11.

Houpt, K.A. (1984) Treatment of aggression in horses. *Equine Pract.*, **6(6)**, 8–10.

Houpt, K.A. and Hintz, H.F. (1983) Some effects of maternal deprivation on maintenance behaviour, spatial relationships and responses to environmental novelty in foals. *Appl. Anim Ethol.*, **9**, 221–30.

Houpt, K.A., and Olm, D. (1984) Foal rejection: A review of 23 cases. *Equine Pract.*, **6(7)**, 38–40.

Houpt, K.A. and Wolski, T. (1982) *Domestic Animal Behavior for Veterinarians and Animal Scientists*, Iowa State University Press, Ames, Iowa.

Hubrecht, R.C., Serpell, J.A. and Poole, T.B. (1992) Correlates of pen size and housing conditions on the behaviour of kennelled dogs. *Appl. Anim. Behav. Sci.*, **34**, 365–83.

Hucklebridge, F.H., Gamal-el-Din, L. and Brain, P.F. (1981) Social status and the adrenal medulla in the house mouse. *Behav. Neur. Biol.*, **33**, 345–63.

Hughes, B.O. (1975) Spatial preference in the domestic. *Br. Vet. J.*, **131**, 560–64.

Hughes, B.O. and Black, A.J. (1973) The preference of domestic hens for different types of battery cage floor. *Br. Poult. Sci.*, **14**, 615–19.

Hughes, B.O. and Duncan, I.J.H. (1988a) Behavioural needs: can they be explained in terms of motivational models? *Appl. Anim. Behav. Sci.*, **20**, 352–55.

Hughes, B.O. and Duncan, I.J.H. (1988b) The notion of ethological 'need', models of motivation and animal welfare. *Anim. Behav.*, **36**, 1696–707.

Hughes, J., Smith, T.W., Kosterlitz, H.W., Fothergill, L.A., Morgan, B.A. and Morris, H.R. (1975) Identification of two related pentapeptides from the brain with potent opiate agonist activity. *Nature, Lond.*, **258**, 577–79.

Huntingford, F. and Turner, A. (1987) *Animal Conflict*, Chapman & Hall, London.

Hurnik, J.F. (1987) Sexual behaviour of female domestic mammals, in *The Veterinary Clinics of North America*, 3, 2, Farm Animal Behavior, (ed E.O. Price), Saunders, Philadelphia, pp. 423–61.

Hurnik, J.F. and Lehman, H. (1988) Ethics and farm animal welfare. *J. Agric. Ethics*, **1**, 305–18.

Hutson, G.D. (1989) Operant tests of access to earth as a reinforcement for weaner piglets. *Anim. Prod.*, **48**, 561–69.

Hutson, G.D. (1992) A comparison of operant responding by farrowing sows for food and nest-building materials. *Appl. Anim. Behav. Sci.*, **34**, 221–30.

Hutson, G.D. and Haskell, M.J. (1990) The behaviour of farrowing sows with free and operant access to an earth floor. *Appl. Anim. Behav. Sci.*, **26**, 363–72.

Hutt, C. and Hutt, S.J. (1970) Stereotypies and their relation to arousal: a study of autistic children, in *Behaviour Studies in Psychiatry*, (eds C. Hutt and S.J. Hutt), Pergamon Press, Oxford, pp. 175–200.

Irwin, M., Daniels, M. and Weiner, H. (1987) Immune and neuroendocrine changes during bereavement. *Psychiat. Clin. North Am.*, **10**, 449–65.

Janssens, L.A.A., Rogers, P.A.M. and Schoen, A.M. (1988) Acupuncture analgesia: review. *Vet. Rec.*, **122**, 355–58.

Jensen, P. (1979) Sinsuggors beteendemonster under tre olika upstallnings forhallanden – en pilot studie. *Institutionen for husdjurshygien med horslarskalan, Rapport 1*, Sveriges Lantbruksuniversitet, Uppsala, pp. 1–40.

Jensen, P. (1980) An ethogram of social interaction patterns in group-housed dry sows. *Appl. Anim. Ethol.*, **6**, 341–50.

Jensen, P. (1986) Observations on the maternal behaviour of free ranging domestic pigs. *Appl. Anim. Behav. Sci.*, **16**, 131–42.

Jensen, P. (1989) Nest site choice and nest-building of free-ranging domestic pigs due to farrow. *Appl. Anim. Behav. Sci.*, **22**, 13–21.

Jensen, P. and Wood-Gush, D.G.M. (1984) Social interactions in a group of free-ranging sows. *Appl. Anim. Behav. Sci.*, **12**, 327–37.

Jephcott, E.H., McMillen, I.C., Rushen, J.P. and Thorburn, G.D. (1987) A comparison of the effects of electroimmobilisation and, or, shearing procedures on ovine plasma concentrations of ß-endorphin/ß-lipoprotein and cortisol. *Res. Vet. Sci.*, **43**, 97–100.

Jeppesen, L.L. and Heller, K.E. (1986) Stress effects on circulating eosinophil leukocytes, breeding performance and reproductive success of ranch mink. *Scientifur*, **10**, 15–18.

Johnson, K.G. and Cabanac, M. (1982) Homeostatic competition between food intake and temperature regulation in rats. *Physiol. Behav.*, **28**, 675–79.

Johnson, K.G. and Hales, J.R.S. (1984) An introductory analysis of competition between thermoregulation and other homeostatic systems, in *Thermal Physiology*, (ed J.R.S. Hales), Raven Press, New York, pp. 295–98.

Johnson, M.D., Shier, D.N. and Barger, A.C. (1979) Circulating catecholamines and control of plasma renin activity in conscious dogs. *Am. J. Physiol.*, **236**, H463–H470.

Jones, A.R. and Price, S. (1990) Can stress in deer be measured? *Deer*, **8**, 25–27.

Kagan, A.R. and Levi, L. (1974) Health and environment – psychosocial stimuli: a review. *Soc. Sci. Med.*, **8**, 225–41.

Kalin, N.H., Carnes, M., Barksdale, C.M., Shelton, S.E., Stewart, R.D. and Risch, S.C. (1985a) Effects of acute behavioural stress on plasma and cerebrospinal fluid ACTH and β-endorphin in rhesus monkeys. *Neuroendocrinol.*, **40**, 97–101.

Kalin, N.H., Gibbs, D.M., Barksdale, C.M. Shelton, S.D. and Carnes, M. (1985b) Behavioral stress decreases plasma oxytocin concentrations in primates. *Life Sci.*, **36**, 1275–80.

Kalin, N.H., Shelton, S.E., Barksdale, C.M. and Carnes, M. (1985c) The diurnal variation of immunoreactive adrenocorticotropin in rhesus monkey plasma and cerebrospinal fluid. *Life Sci.*, **36**, 1135–40.

Kant, G.J., Bunnell, B.N., Mougey, E.H. Pennington, L.L. and Meyerhoff, J.L. (1983a) Effects of repeated stress on pituitary cyclic AMP, and plasma prolactin, corticosterone and growth hormone in male rats. *Pharmacol. Biochem. Behav.*, **18**, 967–71.

Kant, G.J., Leu, J.R., Anderson, S.M. and Mougey, E.H. (1987) Effects of chronic stress on plasma corticosterone, ACTH and prolactin. *Physiol. Behav.*, **40**, 775–79.

Kant, G.J., Mougey, E.H. and Meyerhoff, J.L. (1986) Diurnal variation in neuroendocrine response to stress in rats: plasma ACTH, beta-endorphin, beta-LPH, corticosterone, prolactin and pituitary cyclic AMP responses. *Neuroendocrinol.*, **43**, 383–90.

Kant, G.J., Mougey, E.H., Pennington, L.L. and Meyerhoff, J.L. (1983b) Graded footshock stress elevates pituitary cyclic AMP and plasma β-endorphin, β-LPH, corticosterone and prolactin. *Life Sci.*, **33**, 2657–63.

Keiper, R.R. (1970) Studies of stereotypy function in the canary (*Serinus canarius*). *Anim. Behav.*, **18**, 353–57.

Kelley, K.W. (1980) Stress and immune function. A bibliographic review. *Ann. Rech. Vét.*, **11**, 445–78.

Kelley, K.W. (1985) Immunological consequences of changing environmental stimuli, in *Animal Stress*, (ed G.P. Moberg), American Physiological Association, Bethesda, Maryland, pp. 193–223.

Kelley, K.W., Greenfield, R.E. and Evermann, J.F. (1982) Delayed-type hypersensitivity, contact sensitivity, and phytohemagglutinin skin-test responses of heat- and cold-stressed calves. *Am. J. Vet. Res.*, **43**, 775–79.

Kent, J.E. and Ewbank, R. (1983) Changes in the behaviour of cattle during and after road transportation. *Appl. Anim. Ethol.*, **11**, 85.

Kent, J.E. and Ewbank, R. (1986) The effect of road transportation on the blood constituents and behaviour of calves. *Br. Vet. J.*, **142**, 326–35.

Kent, J.E., Molony, V. and Robertson, I.S. (1991) Responses of different aged lambs to three methods of castration and tail docking, AVTRW conference, Scarborough (unpublished).

Keogh, R.G. and Lynch, J.J. (1982) Early feeding experience and subsequent acceptance of feed by sheep. *Proc. NZ Soc. Anim. Prod.*, **42**, 73–75.

Kerby, C. and MacDonald, D.W. (1988) Cat society and the consequences of colony size, in *The Domestic Cat: the biology of its behaviour*, (eds D.C. Turner and P. Bateson), Cambridge University Press, Cambridge, pp. 67–81.

Kilgour, R. and Dalton, C. (1984) *Livestock Behaviour: a practical guide*, Granada, London.

Komisaruk, B.R. and Larsson, K. (1971) Suppression of a spinal and cranial nerve reflex by vaginal or rectal probing in rats. *Brain Res.*, **35**, 231–35.

Konarska, M., Stewart, R.E. and McCarty, R. (1989a) Sensitization of sympathetic-adrenal medullary responses to a novel stressor in chronically stressed laboratory rats. *Physiol. Behav.*, **46**, 129–35.

Konarska, M., Stewart, R.E. and McCarty, R. (1989b) Habituation of sympathetic-adrenal medullary responses following exposure to chronic intermittent stress. *Physiol. Behav.*, **45**, 255–61.

Koolhaas, J.M., Schuurmann, T. and Fokkema, D.S. (1983) Social behaviour of rats as a model for the psychophysiology of hypertension, in *Biobehavioral bases of coronary heart disease*, (eds T.M. Dembrowski, T.H. Schmidt and G. Blumchen), Karger, Basel, pp. 391–400.

Krulich, L., Hefco, E., Illner, P. and Read, C.B. (1974) The effects of acute stress on the secretion of LH, FSH, prolactin and growth hormone in the normal male rat, with comments on their statistical evaluation. *Neuroendocrinol.*, **16**, 293–311.

Kvetnansky, R., Sun, C.L., Lake, C.R., Thoa, N., Torda, T. and Kopin, I. J. (1978) Effect of handling and forced immobilization on rat plasma levels of epinephrine, norepinephrine and dopamine-b-hydroxylase. *Endocrinology*, **103**, 1868–74.

Kvetnansky, R., Weise, V.K. and Kopin, I.J. (1970) Elevation of adrenal tyrosine hydroxylase and phenylethanolamine-N-methyl transferase by repeated immobilization of rats. *Endocrinology*, **87**, 744–49.

Ladewig, J. (1984) The effect of behavioural stl on the episodic release and circadian variation cortisol in bulls, in *Proc. Int. Cong. Appl. Ethol. Farm Anim.* (eds J. Unshelm, G. van Putten and Zeeb). KTBL, Darmstadt, pp. 339–42.

Ladewig, J. and Smidt, D. (1989) Behaviour, episodic secretion of cortisol, and adrenocortical reactivity in bulls subjected to tethering. *Horm. Behav.*, **23**, 344–60.

Lagadic, H. and Faure, J-M. (1983) Preferences of domestic hens for cage size and floor types as measured by operant conditioning. *Appl. Anim. Behav. Sci.*, **19**, 147–55.

Lagadic, H. and Faure, J.-M. (1987) Preferences of domestic hens for cage size and floor types as measured by operant conditioning. *Appl. Anim. Behav. Sci.*, **19**, 147–55.

Lamprecht, F., Williams, R.B. and Kopin, I.J. (1973) Serum dopamine-beta-hydroxylase during development of immobilisation-induced hypertension. *Endocrinology*, **92**, 953–56.

Larkin, S. and McFarland, D. (1978) The cost of changing from one activity to another. *Anim. Behav.*, **26**, 1237–46.

Laudenslager, M.L., Reite, M. and Harbeck, R.J. (1982) Suppressed immune response in infant monkeys associated with maternal separation. *Behav. Neur. Biol.*, **36**, 40–48.

Lawrence, A.B. and Illius, A.W. (1989) Methodology for measuring hunger and food needs using operant conditioning in pig. *Appl. Anim. Behav. Sci.*, **24**, 273–85.

Lazarus, R.S. and Folkman, S. (1984) *Stress, Appraisal and Coping*, Springer, New York.

Lea, S.E.G. (1978) The Psychology and Economics of Demand. *Psychol. Bull.*, **85**, 441–66.

Lee, D.H.K. (1966) The role of attitude in response to environmental stress. *J. Soc. Issues*, **22**, 83–91.

Lendfers, L.H.H.M. (1970) De invloed van transport op sterfte en vleeskwaliteit van slachtvarkens. *Tijdschr. Diergeneesk.*, **95(25)**: 1331–42.

Levine, S., Goldman, L. and Coover, G.D. (1972) Expectancy and the pituitary-adrenal system, in *Physiology, Emotion and Psychosomatic Illness*, (eds R. Porter and J. Knight), Elsevier, Amsterdam.

Levy, D.M. (1944) On the problem of movement restraint. *Am. J. Orthopsychiat.*, **14**, 644–71.

Lewis, J.W., Terman, G.W., Nelson, L.R. and Liebeskind, J.C. (1984) Opioid and non-opioid stress analgesia, in *Stress-Induced Analgesia*, (eds N.D. Tricklebank and G. Curzon), Wiley, Chichester, pp. 103–33.

Liang, B., Verrier, R.L., Melman, J. and Lown, B. (1979) Correlation between circulating catecholamine levels and ventricular vulnerability during psychological stress in conscious dogs. *Proc. Soc. Exp. Biol. Med.*, **161**, 266–69.

Lindsay, D.R. (1985) Reproductive anomalies, in *Ethology of Farm Animals*, World Animal Science A5, (ed A.F. Fraser), Elsevier, Amsterdam, pp. 413–18.

Livesey, G.T., Miller, J.M. and Vogel, W.H. (1985) Plasma norepinephrine, epinephrine and corticosterone stress responses to restraint in individual male and female rate and their conditions. *Neurosci. Lett.*, **62**, 51–56.

Lockwood, J.A. (1987) The moral standing of insects and the ethics of extinction. *Florida Entomologist*, **70**, 70–89.

MacKay-Sim, A. and Laing, D.G. (1980) Discriminaton of odors from stressed rats by non-stressed rats. *Physiol. Behav.*, **24**, 699–704.

MacLennan, A.J., Drugan, R.C., Hyson, R.L. and Maier, S.F. (1982) Corticosterone: a critical factor in an opioid form of stress-induced analgesia. *Science*, NY, **215**, 1530–32.

McBride, G. (1980) Adaptation and welfare at the man-animal interface, in *Behaviour in Relation to Reproduction, Management and Welfare of Farm Animals*, (eds Wodzicka-Tomaszewska, M., Eday, T.N. and Lynch, J.J.), University of New England, N.S.W., pp. 195–99.

McBride, G., Parer, I.P. and Foenandez, F. (1969) The social organisation and behaviour of the feral domestic fowl. *Anim. Behav. Monogr.*, **2**, 127–181.

McCarty, R. (1983) Stress, behaviour and experimental hypertension. *Neurosci. Biobehav. Rev.*, **7**, 493–502.

McCune, S. (1992) Temperament and welfare of caged cats. Ph. D. thesis. University of Cambridge.

McFarland, D.J. (1971) *Feedback Mechanisms in Animal Behaviour*, Academic Press, London.

McFarland, D.J. (1985) *Animal Behaviour*, Pitman, London.

McFarland, D.J. and Sibly, R.M. (1975) The behavioural final common path. *Phil. Trans. R. Soc. B*, **270**, 265–93.

McGrath, J.E. (1970) *Social and Psychological Factors in Stress*, Holt, Rinehart and Winston, New York.

Maier, S.F. and Jackson, R.L. (1979) Learned helplessness: all of use were right (and wrong): inescapable shock has multiple effects, in *The Biology of Learning and Motivation*, vol 13, (ed G. Bower), Academic Press, New York.

Maier, S.F., Ryan, S.M., Barksdale, C.M. and Kalin, N.H. (1986) Stressor controllability and the pituitary-adrenal system. *Behav. Neurosci.*, **100**, 669–74.

Maier, S.F. and Seligman, M.E.P. (1976) Learned helplessness: theory and evidence. *J. Exp. Psychol. Gen.*, **105**, 3–46.

Marx, D. and Schuster, H. (1980) Ethologische Wahlversuche mit fruhabgesetzten Ferkeln wahrend der Flatdeckhaltung, 1. Mitteilung: Ergebnisse des ersten Abschnitts der Untersuchungen zur tiergerechten Fussbodengestaltung. *Dtsch. tierärtzl. Wschr.*, **87**, 365–400.

Marx, D. and Schuster, H. (1982) Ethologische Wahlversuche mit fruhabgesetzten ferkeln wahrend Flatderckhaltung, 2. Mitteilung: Ergebnisse des zweiten Abschnitts der Untersuchungen zur tiergerechten Fussbodengestaltung. *Dtsch. tierärztl. Wschr.*, **89**, 313–52.

Marx, D. and Schuster, H. (1984) Ethologische Wahlversuche mit fruhabgesetzten Ferkeln wahrend der Flatdeckhaltung, 3. Mitteilung: Ergebnisse der Untersuchungen zur tiergerechten flachengrosse. *Dtsch. tierärztl. Wschr.*, **31**, 18–22.

Mason, G.J. (1991a) Stereotypies: a critical review. *Anim. Behav.*, **41**, 1015–37.

Mason, G.J. (1991b) Stereotypies and suffering. *Behav. Process.*, **25**, 103–15.

Mason, G.J. (1991c) Individual differences in the stereotypies of caged mink. University of Cambridge, Ph.D. thesis.

Mason, G.J. Forms of stereotypic behaviour, in *Stereotypic Behaviour: fundamentals and applications to animal welfare*, (eds A.B. Lawrence and J. Rushen), CAB, Oxford (in press).

Mason, G.J. and Turner, M.A. Mechanisms involved in the development and control of stereotypies, in *Perspectives in Ethology*, (eds P.H. Klopfer and P.P.G. Bateson) (in press).

Mason, J.W. (1968) A review of psychoendocrine research on the pituitary adrenal cortical system. *Psychosom. Med.*, **30**, 576–607.

Mason, J.W. (1971) A re-evaluation of the concept of 'non-specificity' in stress theory. *J. Psychiat. Res.*, **8**, 323–33.

Mason, J.W. (1975a) Psychoendocrine mechanisms in a general perspective of endocrine integration, in *Emotions – their parameters and measurement*, (ed L. Levi), Raven Press, New York, pp. 143–82.

Mason, J.W. (1975b) Emotion as reflected in patterns of endocrine integration, in *Emotions – their parameters and measurement*, (ed L. Levi), Raven Press, New York, pp. 183–91.

Mason, J.W., Brady, J.V. and Tolliver, G.A. (1968a) Plasma and urinary 17-hydroxycorticosteroid responses to 72-hour avoidance sessions in the monkey. *Psychosom. Med.*, **30**, 608–30.

Mason, J.W., Jones, J.A. and Ricketts, P.T. (1968b) Urinary aldosterone and urine volume responses to 72-hour avoidance sessions in the monkey. *Psychosom. Med.*, **30**, 733–45.

Mason, J.W., Kenion, C.C. and Collins, D.R. (1968c) Urinary testosterone response to 72-hour avoidance sessions in the monkey. *Psychosom. Med.*, **30**, 721–32.

Mason, J.W., Taylor, E.D., Brady, J.V. and Tolliver, G.A. (1968d) Urinary estrone, estradiol, and estriol responses to 72-hour avoidance sessions in the monkey. *Psychosom. Med.*, **30**, 696–709.

Mason, J.W, Tolson, W.W., Robinson, J.A. *et al.* (1968e) Urinary androsterone, etiocholanolone, and dehydroepiandrosterone responses to 72 hour avoidance sessions in the monkey. *Psychosom. Med.*, **30**, 710–20.

Mason, J.W., Wherry, F.E. and Brady, J.V. (1968f) Plasma insulin response to 72 hour avoidance sessions in the monkey. *Psychosom. Med.*, **30**, 746–59.

Mason, J.W., Wool, M.S. and Wherry, F.E. (1968g) Plasma growth hormone response to avoidance sessions in the monkey. *Psychosom. Med.*, **30**, 760–73.

Mellor, D.J. and Murray, L. (1989) Effects of tail docking and castration on behaviour and plasma cortisol concentrations in young lambs. *Res. Vet. Sci.*, **46**, 387–91.

Mendl, M.T. (1990) Developmental experience and the potential for suffering: does 'out of experience' mean 'out of mind'? *Behav. Brain Sci.*, **13**, 28–29.

Mendl, M.T. (1991) Some problems with the concept of a cut off point for determining whether an animal's welfare is at risk. *Appl. Anim. Behav. Sci.*, **31**, 139–46.

Mendl, M., Zanella, A.J. and Broom, D.M. (1991) Social rank is related to adrenal cortex activity and reproduction in group housed gilts. *Appl. Anim. Behav. Sci.*, **31**, 290.

Mendl, M., Zanella, A.J. and Broom, D.M. (1992) Physiological and reproductive correlates of behavioural strategies in female domestic pigs. *Anim. Behav.*, **44**, 1107–21.

Met, E.M. de and Halaris, A.E. (1979) Origin and distribution of 3-methoxyphenylethylene glycol in body fluids. *Biochem. Pharmacol.*, **28**, 3043–50.

Metz, J.H.M. (1975) Time patterns of feeding and rumination in domestic cattle. *Meded. Landbhoogesch. Wageningen*, **75–12**, 1–66.

Metz, J.H.M. and Oosterlee, C.C. (1981) Immunologische und ethologische Kriterien fur die artgemassehaltung von Sauen und Ferkeln, in *Aktuelle Arbeiten zur artgemassen Tierhaltung, KTBL Schrift*, **264**, 39–50. KTBL, Darmstadt.

Meunier-Salaun, M.C., Vantrimponte, M.N., Raab, A. and Dantzer, R. (1987) Effect of floor area restriction upon performance, behaviour and physiology of growing-finishing pigs. *J. Anim. Sci.*, **64**, 1371–77.

Meyerhoff, J.L., Oleshansky, M.A. and Mougey, E.H. (1988) Psychologic stress increases plasma levels of prolactin, cortisol and POMC-derived peptides in man. *Psychosom. Med.*, **50**, 295–303.

Meyer-Holzapfel, M. (1968) Abnormal behaviour in zoo animals, in *Abnormal Behavior in Animals*, (ed M.W. Fox), W.B. Saunders, Philadelphia, pp. 476–503.

Mickwitz, G. von (1982) Various transport conditions and their influence on physiological reaction, in *Transport of Animals Intended for Breeding, Production and Slaughter*, (ed R. Moss), Martinus Nijhoff, The Hague, pp. 45–53.

Milenkovic, L., Bogic, L., Musicki, B. and Martinovic, J.V. (1984) Effects of aging on prolactin release after stress in female and male rats. *Acta Endocrinol.*, **107**, 337–39.

Millam, J.R. (1987) Preference of turkey hens for nest-boxes of different levels of interior illumination. *Appl. Anim. Behav. Sci.*, **18**, 341–48.

Millan, M.J., Przewlocki, R., Jerlicz, M., Gromsch, C., Holtt, V. and Herz, A. (1981) Stress-induced release of brain and pituitary β-endorphin: major role of endorphins in generation of hyperthermia, not analgesia. *Brain Res.*, **208**, 325–38.

Miller, H.B. and Williams, W.H. (1983) *Ethics and Animals*, Humana Press, Clifton, New Jersey.

Moberg, G.P. (1985) Biological response to stress: key to assessment of animal well-being?, in *Animal Stress*, (ed G.P. Moberg), American Physiological Society, Bethesda, Maryland, pp. 27–49.

Moberg, G.P. (1987a) Problems in defining stress and distress in animals. *J. Am. Vet. Med. Ass.*, **191**, 1207–11.

Moberg, G.P. (1987b) A model for assessing the impact of behavioral stress on domestic animals. *J. Anim. Sci.*, **65**, 1228–35.

Monjan, A.A. and Collector, M.I. (1977) Stress-induced modulation of the immune response. *Science*, NY, **196**, 307–308.

Morimoto, B.H. and Koshland, D.E. (1991) Short-term and long-term memory in single cells. *FASEB Journal*, **5**, 2061–67.

Mormède, P., Lemaire, V. and Castanon, N. (1990) Multiple neuroendocrine responses to chronic social stress: interaction between individual characteristics and situational factors. *Physiol. Behav.*, **47**, 1099–105.

Mormède, P., Soissons, J., Bluthé, R. N., Raoult, J., Legarft, G., Levieraux, D. and Dantzer, R. (1982) Effects of transportation on blood serum composition, disease incidence and production traits in young calves. Influence of the journey duration. *Annls. Rech. Vét.*, **13**, 369–84.

Morton, D.B. and Griffiths, P.H.M. (1985) Guidelines on the recognition of pain, distress and discomfort in experimental animals and an hypothesis for assessment. *Vet. Rec.*, **116**, 431–36.

Moss, B.W. and McMurray, C.H. (1979) The effect of the duration and type of stress on some serum enzyme levels in pigs. *Res. Vet. Sci.*, **26**, 1–6.

Mueller, G.P. (1981) Beta-endorphin immunoreactivity in rat plasma: variations in response to different stimuli. *Life Sci.*, **29**, 1669–74.

Murata, H. (1989) Suppression of lymphatic blastogenesis of sera from calves transported by road. *Br. Vet. J.*, **145**, 257–62.

Nabeshima, T., Matsuno, K. and Kameyama, T. (1985) Involvement of the different opioid receptors subtypes in electric shock-induced analgesia and motor suppression in the rat. *Eur. J. Pharmacol.*, **114**, 197–207.

Natelson, B.H., Creighton, D., McCarty, R., Tapp, W.N., Pitman, D. and Ottenweller, J.E. (1987) Adrenal hormone indices of stress in laboratory rats. *Physiol. Behav.*, **39**, 117–25.

Natelson, B.H., Kotchen, T.A., Stokes, P.E. and Wooten, G.F. (1977) Relationship between avoidance-induced arousal and plasma DBH, glucose and renin activity. *Physiol. Behav.*, **18**, 671–77.

Natelson, B.H., Tapp, W.N., Adamus, J.E., Mittler, J.C. and Levin, B.E. (1981) Humoral indices of stress in rats. *Physiol. Behav.*, **26**, 1049–54.

Natoli, E. (1985) Spacing patterns in a colony of urban stray cats (*Felis catus*, L) in the historic centre of Rome. *Appl. Anim. Ethol.*, **14**, 289–304.

Nichol, L. (1974) *L'épopée pastorienne et la medicine veterinaire*, Nichol, Garches.

Novak, M.A. and Drewsen, K.H. (1989) Enriching the lives of captive primates, in *Housing, Care and Psychological Wellbeing of Captive and Laboratory Primates*, (ed E.F. Segal), Noyes Publications, Park Ridge, New Jersey, pp. 162–64.

Nyakas, C., Prins, A.J.A. and Bohus, B. (1990) Age-related alterations in cardiac response to emotional stress: relations to behavioural reactivity in the rat. *Physiol. Behav.*, **47**, 273–80.

O'Neill, P.J. and Kaufman, L.N. (1990) Effects of indwelling arterial catheters or physical restraint on food consumption and growth patterns of rats: advantages of noninvasive blood pressure measurement techniques. *Lab. Anim. Sci.*, **40**, 641–43.

Ödberg, F. (1978) Abnormal behaviours: (stereotypies), *Proc. 1st Wld. Congr. Ethol. Appl. Zootechnics, Madrid*, Industrias Grafices España, Madrid, 475–80.

Odio, M., Goliszek, A., Brodish, A. and Ricardo, M.J. (1986) Impairment of immune function after cessation of long-term chronic stress. *Immunol. Lett.*, **13**, 25–31.

Okimura, T., Ogawa, M., Yamauchi, T. and Sasaki, Y. (1986) Stress and immune responses, 4. Adrenal involvement in the attention of antibody responses in restraint-stressed mice. *Jap. J. Pharmacol.*, **41**, 237–45.

Onaka, T., Hamamura, M. and Yagi, K. (1986) Potentiation of vasopressin secretion by footshocks in rats. *Jap. J. Physiol.*, **36**, 1253–60.

Orgeur, P. and Signoret, J.P. (1984) Sexual play and its functional significance in the domestic sheep (*Ovis aries*). *Physiol. Behav.*, **33**, 111–18.

Overmier, J.B., Patterson, J. and Wielkiewicz, R.M. (1980) Environmental contingencies as sources of stress in animals, in *Coping and Health*, (eds S. Levine and H. Ursin), Plenum Press, New York, pp. 1–38.

Paris, J.M., Lorens, S.A., Van de Kar, L.D., Urban, J.H., Richardson-Morton, K.D. and Bethea, C.L. (1987) A comparison of acute stress paradigms: hormonal responses and hypothalamic serotonin. *Physiol. Behav.*, **39**, 33–43.

Parrot, R.F. (1990a) Central administration of corticotropin releasing factor in the pig: effects on operant feeding, drinking and plasma cortisol. *Physiol. Behav.*, **47**, 519–24.

Parrott, R.F. (1990b) Cortisol release in pigs following peripheral and central administration of ovine and human corticotropin releasing hormone. *Acta Endocrinol.*, **123**, 108–12.

Parrott, R.F. (1990c) Physiological responses to isolation in sheep, in *Social Stress in Domestic Animals*, (eds R. Zayan and R. Dantzer), Kluwer Academic Publishers, Dordrecht, pp. 212–26.

Parrott, R.F., Misson, B.H. and Baldwin, B.A. (1989) Salivary cortisol in pigs following adrenocorticotrophic hormone stimulation: comparison with plasma levels. *Br. Vet. J.*, **145**, 362–66.

Parrott, R.F., Thornton, S.N., Forsling, M.L. and Delaney, C.E. (1987) Endocrine and behavioural factors affecting water balance in sheep subjected to isolation stress. *J. Endocrinol.*, **112**, 305–10.

Pavlidis, N and Chirigos, M. (1980) Stress-induced impairment of macrophage tumoricidal function. *Psychosom. Med.*, **42**, 47–54.

Petherick, J.C., and Rutter, S.M. (1990) Quantifying motivation using a computer-controlled push door. *Appl. Anim. Behav. Sci.*, **27**, 159–67.

Pickering, A.D. (1989a) Environmental stress and the survival of brown trout. *Freshw. Biol.*, **21**, 47–55.

Pickering, A.D. (1989b) Factors affecting the susceptibility of salmonid fish to disease. *Rep. Freshw. Biol. Ass.*, U.K. **57**, 61–80.

Pickering, A.D. and Pottinger, J.G. (1985) Factors influencing blood cortisol levels of brown trout under intensive culture conditions, in *Current Trends in Comparative Endocrinology*, (eds B. Lofts and W.N. Holms), Hong Kong University Press, Hong Kong, pp. 1239–42.

Pitman, D.L., Ottenweller, J.E. and Natelson, B.H. (1990) Effect of stressor intensity on habituation and sensitization of glucocorticoid responses in rats. *Behav. Neurosci.*, **104**, 28–36.

Poll, N.E. van der, Jonge, F. de, Oyen, H.G. van and Pelt, J. van (1982) Aggressive behaviour in rats: effects of winning or losing on subsequent aggressive interactions. *Behav. Proc.*, **7**, 143–55.

Pollock, R.E., Lotzova, E., Stanford, S.D. and Romsdahl, M.M. (1987) Effect of surgical stress on murine natural killer cell cytotoxicity. *J. Immunol.*, **138**, 171–78.

Popper, C.W., Chiueh, C.C. and Kopin, I.J. (1977) Plasma catecholamine concentrations in unanaesthetized rats during sleep, wakefulness, immobilization and after decapitation. *J. Pharm. Exp. Ther.*, **202**, 144–48.

Potter, M.J. (1987) Heart Rate and Behaviour in the Domestic Chick. Ph.D. thesis, University of Reading.

Price, E.O. (1985a) Sexual behaviour of large domestic farm animals: an overview. *J. Anim. Sci.*, **61**, Suppl. 3, 62.

Price, E.O. (1985b) Evolutionary and ontogenetic determinants of animal suffering and well-being, in *Animal Stress*, (ed G.P. Moberg), American Physiological Society, Bethesda, Maryland, pp. 15–26.

Putten, G. van (1980) Objective observations on the behaviour of fattening pigs. *Anim. Regul. Stud.*, **3**, 105–18.

Putten, G. van and Dammers, J. (1976) A comparative study of the well-being of piglets reared conventionally and in cages. *Appl. Anim. Ethol.*, **2**, 339–56.

Putten, G. van and Elshof, W.J. (1978) Observations on the effect of transport on the well being and lean quality of slaughter pigs. *Anim. Regul. Stud.*, **1**, 247–71.

Quirce, C.M. and Maickel, R.P. (1981) Alterations of biochemical parameters by acute and repetitive stress situations in mice. *Psychoneuroendocrinol.*, **6**, 91–97.

Raab, A. and Storz, H. (1976) A long term study on the impact of sociopsychic stress in tree-shrews (*Tupaia belangeri*) on central and peripheral tyrosine hydroxylase activity. *J. Comp. Physiol.*, **108**, 115–31.

Rabin, B.S., Lyte, M., Epstein, L.H. and Caggiula, A.R. (1987) Alteration of immune competency by number of mice housed per cage. *Ann. NY Acad. Sci.*, **496**, 492–500.

Regan, T. (1983) *The Case for Animal Rights*, University of California Press, Berkeley, California.

Reite, M., Short, R., Seiler, C. and Pauley, J.D. (1981) Attachment, loss and depression. *J. Child Psychol. Psychiat.*, **22**, 141–69.

Restrepo, C. and Armario, A. (1987) Chronic stress alters pituitary-adrenal function in prepubertal male rats. *Psychoneuroendocrinol.*, **12**, 393–98.

Riad-Fahmy, D., Read, G.F., Walker, R.F. and Griffiths, K. (1982) Steroids in saliva for assessing endocrine function. *Endocr. Rev.*, **3**, 367–95.

Riegle, G.D. and Meites, J. (1976) The effect of stress on serum prolactin in the female rat. *Proc. Soc. Exp. Biol. Med.*, **152**, 441–48.

Riley, V. (1975) Mouse mammary tumors: alteration of incidence as apparent function of stress. *Science*, NY, **189**, 465–67.

Riley, V. (1981) Psychoneuroendocrine influences on immunocompetence and neoplasia. *Science*, **212**, 1100–109.

Robert, S., Matte, J.J., Girard, C.L. *et al.* (1992) Influence of bulky feed on behavior and reproductive performance of gilts and sows. *J. Anim. Sci.*, **70**, Suppl. 1, 158.

Robin, Y. (1986) Aspects symptomatologiques de la douleur chez le chien. *Rec. Méd. Vét.*, **162**, 1333.

Rollin, B. (1981) *Animal Rights and Human Morality*, Prometheus, Buffalo, New York.

Rooijen, J. van (1980) Wahlversuche, eine ethologische Methode zum Sammeln von Messwerten, un Haltungseinflusse zu erfassen und zu beurteilen. *Aktuelle Arbeiten zur artgemassen Tierhaltung, KTBL- Schrift*, **264**, 165–85.

Rooijen, J. van (1981) Die Anpassungsfahigkeit von Schweinen an einstreulose Buchten. *Aktuelle Arbeiten zur artgemassen Tierhaltung, KTBL- Schrift*, **281**, 174–85.

Rosencrans, J.A., Waitzman, N. and Buckley, J.P. (1966) The production of hypertension in mild albino rats subjected to experimental stress. *Biochem. Pharmacol.*, **15**, 1707–18.

Rossier, J., French, E.D., Rivier, C., Ling, N., Guillemin, R. and Bloom, F.E. (1977) Foot-shock induced stress increases β-endorphin levels in blood but not brain. *Nature, Lond.*, **270**, 618–20.

Roth, K.A., Mefford, I.M. and Barchas, J.D. (1982) Epinephrine, norepinephrine, dopamine and serotonin: differential effects of acute and chronic stress on regional brain amines. *Brain Res.*, **239**, 417–24.

Rothschild, M. (1986) *Animals and Man*, Clarendon Press, Oxford.

Rovainen, C.M. and Yan, Q. (1985) Sensory responses of dorsal cells in the lamprey brain. *J. Comp. Physiol.* A., **156**, 181–83.

Rozin, P. (1976) The selection of foods by rats, humans and other animals. *Adr. Stud. Behav.*, **6**, 21–76.

Rubin, R.T., Miller, R.G., Clark, B.R. Poland, R.E. and Arthur, R.J. (1970), The stress of aircraft landings, II. 3-methoxy-4-hydroxyphenylglycol excretion in naval aviators. *Psychosom. Med.*, **32**, 589–97.

Rushen, J. (1984) Stereotyped behaviour adjunctive drinking and the feeding periods of tethered sows. *Anim. Behav.*, **32**, 1059–67.

Rushen, J. (1986a) The validity of behavioural measures of aversion: a review. *Appl. Anim. Behav. Sci.*, **16**, 309–23.

Rushen, J. (1986b) Aversion of sheep for handling treatments: paired choice experiments. *Appl. Anim. Behav. Sci.*, **16**, 363–70.

Rushen, J. (1986c) Some problems with the physiological concept of 'stress'. *Aust. Vet. J.*, **63**, 359–60.

Rushen, J. (1990) Use of aversion-learning techniques to measure distress in sheep. *Appl. Anim. Behav. Sci.*, **28**, 3–14.

Rushen, J (1991) Problems associated with the interpretation of physiological data in the assessment of animal welfare. *Appl. Anim. Behav. Sci.*, **28**, 381–86.

Sachser, N. (1987) Short-term responses of plasma norepinephrine, epinephrine, glucocorticoid and testosterone titers to social and non-social stressors in male guinea pigs of different social status. *Physiol. Behav.*, **39**, 11–20.

Sachser, N. and Lick, C. (1989) Social stress in guinea pigs. *Physiol. Behav.*, **46**, 137–44.

Sainsbury, D.W.B. (1974) *Proceedings 1st International Livestock – Environment Symposium*, 4, American Society of Agricultural Engineers, St Joseph, Missouri.

Sakellaris, P.C. and Vernikos-Danellis, J. (1975) Increased rate of response of the pituitary-adrenal system in rats adapted to chronic stress. *Endocrinology*, **97**, 597–602.

Sambraus, H.H. (1976) Kronismus bein Schweinen. *Dtsch. tierarzlt. Wschr.*, **83**, 17–19.

Sandman, C.A., Barron, J.L., Chicz-De Met, A. and De Met, E.M. (1990) Plasma endorphin levels in patients with self-injurious behaviour and stereotypy. *Am. J. Ment. Retard.*, **95**, 84–92.

Sanford, J., Ewbank, R., Molony, V., Tavernor, W.D. and Uvarov, O. (1986) Guidelines for the recognition and assessment of pain in animals. *Vet. Rec.*, **118**, 334–38.

Sapolsky, R.M. (1983) Individual differences in cortisol secretory patterns in the wild baboon: role of negative feedback sensitivity. *Endocrinology,* **113**, 2263–67.

Sassenrath, E.N. (1970) Increased adrenal responsiveness related to social stress in rhesus monkeys. *Horm. Behav.*, **1**, 283–98.

Schleifer, S.J., Keller, S.E., Camerino, M. *et al.* (1983) Suppression of lymphocyte stimulation following bereavement. *J. Am. Med. Ass.*, **250**, 374–77.

Schmidt, M. (1982) Abnormal oral behaviour in pigs, in *Disturbed Behaviour in Farm Animals*, (ed W. Bessei, *Hohenheimer Arbeiten*, **121**, 115–121, Eugen, Ulmer, Stuttgart.

Schmidt, R.F. and Thews, G. (eds) (1983) *Human Physiology*, Springer-Verlag, Berlin, p. 670.

Schouten, W.G.P. (1986) Rearing conditions and behaviour in pigs. Ph.D. thesis, Agricultural University of Wageningen.

Sedlock, M.L., Ravitch, J. and Edwards, D.J. (1985) The effect of dietary precursors on the excretion of amines and their metabolites in the rat. *Biochem. Med.*, **34**, 318–26.

Seggie, J.A. and Brown, G.M. (1975) Stress response patterns of plasma corticosterone, prolactin, and growth hormone in the rat, following handling or exposure to novel environment. *Can. J. Physiol. Pharmacol.*, **53**, 629–37.

Selye, H. (1950) *The Physiology and Pathology of Exposure to Stress,* Acta, Montreal.

Selye, H. (1973) The evolution of the stress concept. *Amer. Scient.*, **61**, 692–99.

Selye, H. (1976) *The Stress of Life*, 2nd edn, McGraw-Hill Book Co., New York.

Serpell, J. (1986) *In the Company of Animals*, Blackwell, Oxford, p. 215.

Sherwin, C.M. and Johnson, K.G. (1987) The influence of social factors on the use of shade by sheep. *Appl. Anim. Behav. Sci.*, **18**, 143–55.

Shively, C. and Kaplan, J. (1984) Effects of social factors on adrenal weight and related physiology of *Macaca fascicularis*. *Physiol. Behav.*, **33**, 777–82.

Shorten, M. (1954) The reaction of the brown rat towards changes in its environment, in *Control of Rats and Mice*, vol. 2, (ed D. Chitty), Oxford University Press, Oxford.

Shutt, D.A., Fell, L.R., Cornell, R. *et al.* (1987) Stress induced changes in plasma concentrations of immunoreactive endorphin and cortisol in response to routine surgical procedures in lambs. *Aust. J. Biol. Sci.*, **40**, 97–103.

Sibly, R. (1975) How incentive and deficit determine feeding tendency. *Anim. Behav.*, **23**, 437–46.

Sibly, R. and Calow, P. (1983) An integrated approach to life-cycle evolution using selective landscapes. *J. Theor. Biol.*, **102**, 527–47.

Sibly, R. and McCleery, R.H. (1976) The dominance boundary method of determining motivational state. *Anim. Behav.*, **24**, 108–24.

Sibly, R. and McFarland, D. (1974) A state-space approach to motivation, in *Motivational Control Systems Analysis*, (ed D.J. McFarland), Academic Press, London.

Siegel, H.S. (1987) Effects of behavioural and physical stressors on immune responses, in *Biology of Stress in Farm Animals*, (eds P.R. Wiepkema and P.W.M. van Adrichem), *Curr. Top. Vet. Med. Anim. Sci.*, Martinus Nijhoff, Dordrecht, pp. 39–54.

Silver, G.V. and Price, E.O. (1986) Effects of individual vs. group-rearing on the sexual behaviour of pre-puberal beef bulls: mount orientation and sexual responsiveness. *Appl. Anim. Behav. Sci.*, **15**, 287.

Silver, I.A. (1982) *The firing of Horses. Final report to the Veterinary Advisory Committee of the Horserace Betting Levy Board (1982)*, p. 48.
Simon, E. (ed) (1987) Glossary of terms for thermal physiology. *Pflügers Arch.*, **410**, 567–87.
Singer, P. (1990) *Animal Liberation*, 2nd edn, Jonathan Cape, London.
Smelik, P.G. (1987) Adaptation and brain function. *Prog. Brain. Res.*, **72**, 3–9.
Smith, R., Besser, G.M. and Rees, L.H. (1985) The effect of surgery on plasma β-endorphin and methionine-enkephalin. *Neurosci. Lett.*, **55**, 17–21.
Sokolov, E.N. (1960) Neuronal models and the orienting reflex, in *The Central Nervous System and Behavior*, (ed M.A. Brazier), Macy Foundation, New York.
Souza, E.B. de and Loon, G.R. van (1985) Differential plasma β-endorphin, β-lipotropin and adrenocorticotropin responses to stress in rats. *Endocrinology*, **116**, 1577–86.
Spielberger, C.D. (1966) *Anxiety and Behavior*, Academic Press, New York.
Spielberger, C.D. (ed) (1972) *Anxiety: current trends in theory and research*, Academic Press, New York.
Spinelli, J.S. and Markowitz, H. (1987) Clinical recognition and anticipation of situations likely to induce suffering in animals. *J. Am. Vet. Med. Ass.*, **191**, 1216–18.
Staddon, J.E.R. (1980) Optimality analysis of operant behaviour and their relations to optional foraging, in *Limits to Action: the allocation of individual behaviour*, (ed J.E.R. Staddon), Academic Press, New York.
Steenbergen, J.M., Koolhaas, J.M., Strubbe, J.H. and Bohus, B. (1989) Behavioral and cardiac responses to a sudden change in environmental stimuli: effect of forced shift in food intake. *Physiol. Behav.*, **45**, 729–33.
Stein, M., Keller, S.E. and Schleifer, S.J. (1985) Stress and immunomodulation: the role of depression and neuroendocrine function. *J. Immunol.*, **135**, 827S–833S.
Steinberg, H. and Watson, R.H.J. (1960) Failure of growth in disturbed laboratory rats. *Nature, Lond.*, **185**, 615–16.
Stephens, D.B. (1980) Stress and its measurement in domestic animals: a review of behavioural and physiological studies under field and laboratory conditions. *Adv. Vet. Soc. Comp. Med.*, **24**, 179–210.
Stephens, D.B. (1988) A review of experimental approaches to the analysis of emotional behaviour and their relation to stress in farm animals. *Cornell Vet.*, **78**, 155–77.
Stephens, D.B. and Toner, J.N. (1975) Husbandry influences on some physiological parameters of emotional responses in calves. *Appl. Anim. Ethol.*, **1**, 233–43.
Steplewski, Z., Goldman, P.R. and Vogel, W.H. (1987) Effect of housing stress on the formation and development of tumors in rats. *Cancer Lett.*, **34**, 257–61.
Steplewski, Z. and Vogel, W.H. (1986) Total leukocytes, T cell subpopulation and Natural Killer (NK) cell activity in rats exposed to restraint stress. *Life Sci.*, **38**, 2419–27.
Steyn, D.C. (1975) The effects of captivity stress on the blood chemical values of the chacma baboon (*Papio ursinus*). *Lab. Anim.*, **9**, 111–20.
Stolba, A. (1982) A family system of pig housing, in *Proc. Symp. Alternatives to Intensive Husbandry Systems*, Universities Federation for Animal Welfare, Potters Bar.

Stolba, A., Baker, N. and Wood-Gush, D.G.M. (1983) The characterisation of stereotyped behaviour in stalled sows by informational redundancy. *Behaviour*, **87**, 157–82

Stolba, A. and Wood-Gush, D.G.M. (1989) The behaviour of pigs in a semi-natural environment. *Anim. Prod.*, **48**, 419–25.

Stott, G.H. (1981) What is animal stress and how is it measured? *J. Anim. Sci.*, **52**, 519–57.

Swenson, R.M. and Vogel, W.H. (1983) Plasma catecholamine and corticosterone as well as brain catecholamine changes during coping in rats exposed to stressful footshock. *Pharmacol. Biochem. Behav.*, **18**, 689–93.

Syme, L.H. and Elphick, G.R. (1982) Heart-rate and the behaviour of sheep in yards. *Appl. Anim. Ethol.*, **9**, 31–35.

Taché, Y., Ruisseau, P. du, Ducharme, J.R. and Collu, R. (1978) Pattern of adenohypophyseal hormone changes in male rats following chronic stress. *Neuroendocrinol.*, **26**, 208–19.

Tarrant, P.V. (1981) The occurrence, causes and economic consequences of dark-cutting in beef – a survey of current information, in *The Problem of Dark-Cutting in Beef*, (eds D.E. Hood and P.V. Tarrant), Martinus Nijhoff, The Hague.

Taylor, P.M. (1987) Some aspects of the stress response to anaesthesia and surgery in the horse. Ph.D. thesis, University of Cambridge.

Thompson, D.L., Elgert, K.D., Gross, W.B. and Siegel, P.B. (1980) Cell mediated immunity in Mareks disease virus-infected chickens genetically selected for high and low concentrations of plasma corticosterone. *Amer. J. Vet. Res.*, **41**, 91–96.

Thornton, S.N., Parrott, R.F. and Delaney, C.E. (1987) Differential responses of plasma oxytocin and vasopressin to dehydration in non-stressed sheep. *Acta Endocrinol.*, **114**, 519–23.

Tilbrook, A.J. and Cameron, A.W.N. (1989) Ram mating preferences for woolly rather than recently shorn ewes. *Appl. Anim. Behav. Sci.*, **24**, 301–12.

Toates, F. (1986) *Motivational Systems*, Cambridge University Press, Cambridge, p. 188.

Toates, F. and Jensen, P. (1991) Ethological and psychological models of motivation: towards a synthesis, in *Farm Animals to Animats*, (eds J.-A. Meyer and S. Wilson), MIT Press, Cambridge, pp. 194–205.

Trumbull, R. and Appley, M.H. (1986) A conceptual model for the examination of stress dynamics, in *Dynamics of Stress: physiological, psychological and social perspectives*, (eds M.H. Appley and R. Trumbull), Plenum Press, New York.

Trunkfield, H.R. (1990) The effects of previous housing experience on calf responses to housing and transport. Ph. D. thesis, University of Cambridge.

Trunkfield, H.R. and Broom, D.M. (1990) The welfare of calves during handling and transport. *Appl. Anim. Behav. Sci.*, **28**, 135–52.

Trunkfield, H.R., Broom, D.M., Maatje, K., Wieranga, H.K., Lambooy, E. and Kooijman, J. (1991) Effects of housing on responses of veal calves to handling and transport, in *New Trends in Veal Calf Production*, (eds J.H.M. Metz and C.M. Groenestein), Pudoc, Wageningen, pp. 40–43.

Turck, K.H. de and Vogel, W.H. (1980) Factors influencing plasma catecholamine levels in rats during immobilization. *Pharmacol. Biochem. Behav.*, **13**, 129–31.

Turkhan, J.S., Ator, N.A., Brady, J.V. and Craven, K.A. (1989) Beyond chronic catheterization in laboratory primates, in *Housing, Care and Psychological*

Wellbeing of Captive and Laboratory Primates, (ed E.F. Segal), Noyes Publications, Park Ridge, New York, pp. 305–22.

Turpen, C., Johnson, D.C. and Dunn, J.D. (1976) Stress-induced gonadotrophin and prolactin secretory pattern. *Neuroendocrinol.*, **20**, 339–51.

Ullberg, M. and Jondal, M. (1981) Recycling target binding capacity of human Natural Killer cells. *J. Exp. Med.*, **153**, 615–28.

Underwood, A.J. (1989) The analysis of stress in natural populations. *Biol. J. Linnean Soc.*, **37**, 51–78.

Verheijen, F.J. and Buwalda, R.J.A. (1988) *Doen pijn en angst een gehaakte en gedrilde karper lijden*, Report of the Department of Comparative Physiology, University of Utrecht.

Vince, M.A. (1966) Artificial acceleration of hatching in quail embryos. *Anim. Behav.*, **14**, 389–94.

Vining, R.F., McGinley, R.A., Maksvytis, J.J. and Ho, K.Y. (1983a) Salivary cortisol: a better measure of adrenal cortical function than serum cortisol. *Ann. Clin. Biochem.*, **20**, 329–35.

Vining, R.F., McGinley, R.A. and Symons, R.G. (1983b) Hormones in saliva: mode of entry and consequent implications for clinical interpretation. *Clin. Chem.*, **29**, 1752–56.

Wallace, J., Sanford, J., Smith, M.W. and Spencer, K.V. (1990) The assessment and control of the severity of scientific procedures on laboratory animals. *Lab. Anim.*, **24**, 97–130.

Warburton, D.M. (1979) Stress and the processing of information, in *Human Stress and Cognition*, (eds V. Hamilton and D.M. Warburton), Wiley, Chichester, pp. 470–75.

Warwick, C. (1989) The welfare of reptiles in captivity, in *Proc. 1st. Wld. Cong.* Herpetol. Health Dis., Canterbury.

Weiss, J.M. (1971) Effects of coping behaviour in different warning signal conditions on stress pathology in rats. *J. Comp. Physiol. Psychol.*, **77**, 1–13.

Weiss, J.M., Stone, E.A. and Harrell, N. (1970) Coping behavior and brain norepinephrine level in rats. *J. Comp. Physiol. Psychol.*, **72**, 153–60.

Wemelsfelder, F. (1990) Boredom and laboratory animal welfare, in *The Experimental Animal in Biomedical Research*, vol. I, (eds B.E. Rollin and M.L. Kesel), CRC Press, Ann Arbor, Boston, pp. 243–72.

Wemelsfelder, F. and Putten, G. van (1985) Behaviour as a possible indicator for pain in piglets, I.V.0. *Report B-260*, Institut voor Veeteelkundig Onderzoek, Zeist.

West, J.B. (ed) (1990) *Best and Taylor's 'Physiological Basis of Medical Practice'*, 12th edn, Williams and Wilkins, Baltimore.

Wiepkema, P.R. (1987) Behavioural aspects of stress, in *Biology of Stress in Farm Animals: an integrative approach*, (eds P.R. Wiepkema and P.W.M. van Adrichem), *Curr. Top. Vet. Med. Anim. Sci.*, Martinus Nijhoff, Dordrecht, pp. 113–83.

Wiepkema, P.R., Broom, D.M., Duncan, I.J.H. and Putten, G. van (1983) *Abnormal Behaviours in Farm Animals*, Commission of the European Communities, Brussels.

Wood, G.N. and Molony, V. (1991) Sensitivity of behaviour and plasma cortisol to changes in acute pain from the testes and scrotum in young lambs. AVTRW conference, Scarborough (unpublished).

Wood-Gush, D.G.M. (1988) The relevance of the knowledge of free ranging domesticated animals for animal husbandry, in *Proc. Int. Cong. Appl. Ethol. Farm Animals,* (eds G. van Putten, J. Unshelm and K. Zeeb), KTBL, Darmstadt.

Wood-Gush, D.G.M. and Beilharz, R.G. (1983) The enrichment of a bare environment for animals in confined conditions. *Appl. Anim. Ethol.*, **10**, 209–17.

Wood-Gush, D.G.M. and Vestergaard, K. (1989) Exploratory behaviour and the welfare of intensively kept animals. *J. Agric. Ethics*, **2**, 161–69.

Wright, E.M. and Woodson, J.F. (1990) Clinical assessment of pain in laboratory animals, in *The Experimental Animal in Biomedical Research*, vol. 1, (eds B.E. Rollin and M.L. Kesel), CRC Press, Boca Raton, Florida, pp. 205–15.

Wuttke, W., Denling, J., Roosen-Runge, G. *et al.* (1984) In vivo release of catecholamines and amino acid neurotransmitters, in *Stress: the role of catecholamines and other neurotransmitters*, vol. 1, (eds E. Usdin, R. Kvetnansky and J. Axelrod), Gordon and Breach Science Publishers, New York, pp. 93–103.

Yousef, M.K. (ed) (1984) Stress physiology: definition and terminology, in *Stress Physiology in Livestock*, vol. 1, *Basic Principles.*, CRC Press, Boca Raton, Florida.

Zanella, A.J. (1992) Sow welfare indicators and their inter-relationships. Ph. D. thesis, University of Cambridge.

Zanella, A.J., Broom, D.M. and Hunter J.C. (1991a) Changes in opioid receptors of sows in relation to housing, inactivity and stereotypies, in *Applied Animal Behaviour: past, present and future*, (eds M.C. Appleby, R.I. Horrell, J.C. Petherick and S.M. Rutter), Universities Federation for Animal Welfare, Potters Bar, pp. 140–41.

Zanella, A.J., Broom, D.M. and Hunter, J.C. (1992) Changes in opioid receptors in sows in relation to housing, inactivity and stereotypies, in *Animal Welfare: proceedings of the XXIV World Veterinary Congress, Rio de Janeiro 1991*, World Veterinary Association, London, pp. 159–66.

Zanella, A.J., Broom, D.M. and Mendl, M.T. (1991b) Responses to housing conditions and immunological state in sows. *Anim. Prod.*, **52**, 579.

Zanella, A.J., Mendl, M., and Broom, D.M. (1991c) An investigation of the relationship between adrenal activity, social rank and immunocompetence in pregnant sows kept in different housing conditions. *Appl. Anim. Behav. Sci.*, **30**, 175–76.

Zeman, P., Alexandrova, M., Kvetnansky, R. (1988) Opioid u delta and dopamine receptor numbers in rat striatum during stress. *Appl. Anim. Behav. Sci.*, **30**, 175–76.

Zenchak, J. and Anderson, C.C. (1980) Sexual performance levels of rams (*Ovis aries*) as affected by social experiences during rearing. *J. Anim. Sci.*, **50**, 167–74.

Zielinsky, D. (1989) Estimation of vanillylmandelic acid in parotid saliva. *J. Clin. Chem. Clin. Biochem.*, **27**, 238–39.

Zimmermann, M. (1985) Behavioral investigations of pain in animals, in *Proc. 2nd. Eur. Symp. Poult. Welfare*, (ed R.-M. Wegner), World Poultry Science Association, Celle.

Zito, C.A., Wilson, L.L. and Graves, H.B. (1977) Some effects of social deprivation on behavioural development of lambs. *Appl. Anim. Ethol.*, **3**, 367–77.

Index

Page numbers in **bold** represent detailed discussions of the entries